Advance praise for *Lessons in Agile Management*

"David J. Anderson's work has proven to be foundational in helping my clients walk the road of continuous improvement. David's latest book contains anecdotes that are guaranteed to educate, enlighten, and even amuse those knowledge workers seeking better organizational agility."

–Jeff Anderson, Service Offering Lead, Deloitte LEAN

"In an anecdotic, analytic, and easy-to-digest way, David explains all the evolving thinking behind such a simple and powerful method [Kanban], from Deming, Goldratt, and beyond. David challenges both conventional and traditional Agile mindsets by thinking in terms of the best economical outcomes, value, flow, and the capability of the system and the humans working as part of it."

Manuel Mazán, MBA, CSP, Lean-Agile Coach, Agiland

"David's writing has inspired many software managers and leaders over the last ten years. These collected and annotated articles provide a fresh perspective on many of his best posts, articles, and insights. I highly recommend Lessons in Agile Management for those who seek an understanding of both the current best practice and the evolution of software management. Neither stands without the other."

Eric Willeke, Agile Coach, Rally Software Development

"I have been following David's work with Kanban for the last 3 years. His work has helped shape the Kanban implementation within our matrix organization. *Lessons in Agile Management: On the Road to Kanban* joins my required reading list on this subject. The plethora of experiences outlined in the book will provide insight and help you avoid the pitfalls as you transform your organization for breakthrough performance. David successfully blends the principles of flow, queues, theory of constraints, Lean, and Six Sigma to concoct a very powerful message that I have come to believe over the years: 'You've got more capacity than you think!'"

Charan Atreya, PMP, Kanban Way

"It's a great book. If the number of pages that now have underlines is an indication of the book's impact, this is a real winner . . . David Anderson ranks in the top few authors whose writing is always a must read. He has done it again with *Lessons in Agile Management*. This latest book brings together diverse bodies of knowledge from queuing theory, cycle times, and flow to the nature of knowledge work, leading knowledge workers, and improving the effectiveness of teams."

Greg Cohen CPM CPMM, Senior Principal Consultant, The 280 Group

"This is the book on leadership we have been waiting for. It is even relevant outside of software development."

Kurt Häusler, CSP, Agile & Lean Consultant, Kegon AG

Lessons in
Agile Management
On the Road to Kanban

David J. Anderson

BLUE
HOLE
PRESS

Sequim, Washington

Blue Hole Press
Sequim, WA 98382
www.blueholepress.com

Printed in the United States of America.

Library of Congress Cataloging-in-Publication Data applied for; available at www.loc.gov

978-0-9853051-2-3   Paperback
978-0-9853051-3-0   Kindle
978-0-9853051-4-7   ePub

Cover design: Terry Kluytmans
Interior design: Vicki Rowland

In loving memory of my father, William E. Anderson

Contents

Foreword
by Alan Shalloway

AFTER 40 YEARS DOING SOFTWARE DEVELOPMENT, I AM CONVINCED THAT HERACLITUS (535 BC–475 BC) WAS CORRECT, "CHANGE IS THE ONLY CONSTANT." BEING IN THIS industry so long doesn't make me smarter; it just means I've seen a lot! And what I have seen is not just change, but the pattern of change. In the early '70s, we debated the merits of interactive programming (using a terminal) over batch programming (cards). A little later in the decade, we were debating the merits of structured programming. And the '70s also gave rise to the Waterfall method. The '80s ushered in micro- and personal computers, and the ability to break the hold that IT organizations had on innovation. Object-orientation became popular. The '90s brought Agile to the fore.

It makes me think about Arthur Schopenauer's quote:

> All truth passes through three stages. First, it is ridiculed. Second, it is violently opposed. Third, it is accepted as being self-evident.

Perhaps you would argue that not all of these changes were good. Waterfall comes to mind. But I would suggest that each change brought to light something that *was* good. Waterfall brought up concerns that needed to be addressed, even if its approach was not a good solution.

Here is what I think: The pattern of change is multifold, and it includes the following:

- ❏ Change brings new ideas.

- ❏ There is always going to be something better.

- ❏ Those who bring the initial change often resist the next wave of change.

- ❏ The acceptance of new ideas takes time.

- ❏ There is overlap between one stage and the next.

We talk about "internet time," meaning that learning has accelerated since the invention of the internet. I am sure that the internet has been accelerating the rate at which we accept new ideas and, I am sure, we still have a long way to go. Ideas are tricky, however. They are not facts. The most important ideas are those that affect how we see the world. These are called paradigms, or mindsets. We hold on dearly to paradigms, especially when it involves our survival, our image, or our professional livelihood. Changing paradigms or mindsets is hard!

Recently, I've been involved in process and methods—eXtreme Programming, Scrum, Lean, and Kanban. The patterns I saw in my first 30 years, regarding technical issues, I'm seeing now, in the last 15, regarding process and methods. The Agile movement has been a movement of change, a movement based on accelerating change. Unfortunately, this change has come in fits and starts: a series of good leaps forward that then became somewhat entrenched. This is natural; we are only human, and we get attached to our ideas. We've been embracing some ideas, rejecting others, and totally ignoring still others. The good news is that there are some people who are more attached to advancing the state of what is possible than they are to their own personal beliefs at the moment. These are the true thought leaders. David Anderson is one such person—and this book deals with many of the topics that need to be, but have rarely been, delved into deeply in the Agile community—leadership, management, and tribal behavior, to name a few.

Over the past five years, I have seen David's work address many of the issues that we, at Net Objectives, have come across in our consulting work with organizations. We kept bumping up against the limits of various Agile methods as we rolled them out in the real world. We started with eXtreme Programming, moved to Scrum, and now also embrace Lean and Kanban. This was a big task, incorporating the good in these approaches while extending them. David Anderson's work helped us significantly.

Here are some of the most important challenges that his work has addressed:

- ❏ The role of leadership in Agile adoption

- ❏ The role of management in Agile adoption

- ❏ How to manage teams based on the theories of flow

- ❏ How to help management understand a team's workflow

- ❏ How to transition teams to a continuous learning model

- ❏ How to help people learn for themselves
- ❏ How to implement Agile methods when cross-functional teams are difficult to create
- ❏ How to remove the bottlenecks caused by key people being overwhelmed with work (sometimes this is the entire development team)

David's earlier book, *Kanban: Successful Evolutionary Change for Your Technology Business*, provided great insights both on transforming business and on implementing Kanban.

When David told me about this new book and asked me to write a foreword, I was both honored and happy to say yes. I had expected some deeper insights, perhaps extensions to his other works. As with most great books, I got something different from what I expected.

I have long been saying that we now know mostly what to do in building software. The real challenge is in getting people to do it. As Gerry Weinberg has said for decades, "It's not a technical problem, it's a people problem." Software development is complicated; getting people to do it properly is complex.

So, in one sense, this book does not present new insights. Much of it comprises quotes that are decades old. However, David presents these "old" insights in a way that provides new opportunities. He relates his thoughts in the form of stories and anecdotes that are quite entertaining. These stories help us see how smart people evolve their ideas. And they help us access the ideas we need.

I happened to be reading an advance copy of this book while at a client engagement. I found myself repeatedly presenting insights from the book. They were very accessible to my clients. I suspect this will be your experience as well. While you may not agree with everything David has to say, I promise that this book will make you think. And it will equip to you convey whatever conclusions you come to much more effectively.

I do not believe the Agile community has yet embraced the role of leadership and management outside the teams. We have not yet fully embraced the impact of tribal behavior in both software development organizations and in those attempting to convey new methods of creating software. This book provides insights into all of these areas.

I think you will find this of value. You will be getting into the head of one of the most thoughtful and influential Agile thought leaders of our time. David's consistently out-of-the-box thinking will help you get out of your own box, no matter what that is.

Foreword

by Stephen Denning

O F ALL THE MANY SINS OF CURRENT MANAGEMENT PRAC-
TICE, ONE OF THE MOST TROUBLING IS ITS FAILURE TO
ADVANCE. IDEAS THAT ARE PRESENTED AS "THE NEXT REALLY
big new thing" have often been around for decades. It is as
though business concepts experience a series of "virgin births":
they come into the world, apparently without any legitimate
parentage, only to be set aside in favor of another new idea, to
be reborn again a few years later with some new label. The tendency to forget the past and to present old ideas with shrieks of
"Eureka!" is not only misleading: It prevents progress. The failure
to present a coherent perspective of where ideas have come from,
and the absence of any diagnosis as to why seemingly good ideas
were set aside in the past, hampers both the identification of
root causes and the articulation of any coherent vision of where
we might be heading or what we should do next. Lessons once
learned end up being relearned over and over again. Often, the
result is constant churning rather than real forward progress.

A welcome exception to these tendencies is David
Anderson's *Lessons in Agile Management: On the Road to
Kanban.* The book is a compendium of blog posts written over
a 13-year period. The separate pieces have been curated into a
set of themes, and the themes woven together provide an overall perspective of how software is developed, including per-
spectives on leadership, management, communities, risk, roles
and responsibilities, human resources, Agile, Kanban, Lean,
and more.

The result reflects the thinking that flowed from *Agile
Management for Software Engineering: Applying the Theory of
Constraints for Business Results* (2003) and foreshadows what
was to come in *Kanban: Successful Evolutionary Change for
Your Technology Business* (2010).

The book comprises a treasury of insights in the evolution of thinking about software development and Kanban, showing its deep roots and distinguished lineage. Here we learn about the contributions of giants like W. Edwards Deming, Peter Drucker, and Eliyahu Goldratt. The book shows, among many other things, that Agile thinking did not begin in 2001 with the Agile Manifesto. Its foundations go back many decades.

Another key value of this book is its open-mindedness and the absence of dogmatism. The book views Lean, Six Sigma, and the Theory of Constraints as three different lenses for considering process improvements. Each method has its own community. Communities being tribal in nature, tribal membership is recognized by the practices performed. Yet each lens is useful. Lean, with its focus on flow, Six Sigma, with its focus on reducing variation, and the Theory of Constraints, with its focus on identifying constraints and relieving them, are complementary. Viewing a problem through just one of these lenses can be limiting. The book encourages the use of multiple models, which generates more innovative ideas for process improvement.

Although Kanban has often been associated with Lean manufacturing, it has now spawned an umbrella movement in software development, as well as wider applicability in legal work, architecture, web services, sales, marketing, human resources, and finance. The movement provides a home for process-improvement thinkers from the various schools of thought. Happily, the Kanban community welcomes practitioners of the different approaches.

Although the adoption of Kanban in software development occurred after the promulgation of the Agile Manifesto, it has spread quickly into countries around the world. The book suggests that this rapid adoption has occurred in part because Kanban is relatively compatible with the "control culture" that is still prevalent in large organizations today. By contrast, other forms of Agile development tend to be more dependent on collaboration and mutual trust, and they sometimes encounter considerable tension in a control culture.

Thus, Kanban can flourish even in organizations with a conservative, low-trust, controlling approach to management, corporate governance, and investment. For those working in such environments, Kanban can be a good fit because it doesn't challenge the prevailing culture.

David Anderson is to be congratulated for having rescued these cool, calm, thoughtful reflections and made them conveniently available in this volume. Together they represent a helpful set of insights on what makes software development really work in today's workplace.

—Stephen Denning, author of *The Leader's Guide To Radical Management: Reinventing the Workplace for the 21st Century* (Jossey-Bass, 2010).

Preface

THIS BOOK WAS CONCEIVED AS "THE *Very* BEST OF AGILE MANAGEMENT BLOG." IN PART AS A WAY TO BRING CLOSURE TO THE AGILEMANAGEMENT.NET WEB SITE, AS MY own business transitions to our new domains, djaa.com and leankanbanuniversity.com, I wanted to make some of the older material more accessible. I felt there was a lot of value in things I'd written almost a decade ago, and that I still actively use in my training and consulting practice. It would be easy, then, to dismiss this book as merely a compendium of blog posts, all of which are freely available on my web site.

Working with my editor, Janice Linden-Reed, we started the project in the fall of 2011. It quickly became evident that "the *very* best of Agile Management blog" was too large a scope. One of the first jobs was to identify articles we thought were worthy of inclusion, and to classify them. In the end, we decided to drop all the articles written about programming in the Feature-Driven Development (FDD) method. We did retain a few articles on analysis, design, and software architecture, but all the object-modeling, coding, and code organization (configuration management and version control) advice is largely gone. We also chose to omit all the articles relating to politics in the Agile community, even though a number of those articles highly influenced events at the Agile Alliance, the Agile conference, and, to a lesser extent, the wider community. We felt that Agile community politics wasn't relative to this narrative. On a related theme, we chose to omit some historically significant posts such as those relating to the now-famous XIT Sustaining Engineering team case study at Microsoft in 2005. And finally, we didn't include many fine articles about Kanban, as we felt those were adequately covered in my second book, *Kanban:*

Successful Evolutionary Change for Your Technology Business, or will be covered in subsequent books about the Kanban Method.

A theme emerged—this book should define the *missing link*. It should be both the sequel to my first book, *Agile Management for Software Engineering,* and the prequel to *Kanban*. There was an opportunity to curate the blog posts and articles that would explain the thinking process that brought about Kanban, not just as a method for improving predictability and managing change and risk in technology businesses, but as a community and a movement. We felt that this would provide the most value to those struggling to manage and lead in the Agile era and would provide an understanding of why Kanban is such an important innovation for improving the agility and culture at technology businesses.

In all, there are around 110,000 words here. Compare that with 63,500 and 72,000 for my previous two books. Each chapter has a theme that I feel is relevant to managers who must lead in the age of Agile adoption. Each chapter begins with contemporary commentary that puts the chosen articles in context and relates the lessons to both Agile methods and to Lean software development and the Kanban Method. Many of the articles are embellished with additional commentary and updates that enhance the understanding and underscore the lessons. The entire 110,000 words have been read and re-read. Every sentence has been touched and improved for clarity, readability, grammar, and punctuation. My copy editor Vicki Rowland has worked hard to turn my poorly written, overly complicated, crusty old musings into something digestible, and I hope, comprehendible. As such, the diligent reader will find that each article isn't copied verbatim from the site, nor does the content always accurately reflect what was written at the time. Some articles have been edited for clarity, and in a few instances, corrected for accuracy or consistency with other, usually later, articles. The intent has been to teach valuable lessons, not to confuse in the name of historical accuracy. In general, where I learned a lesson and changed my mind about something, I refer to this in the contemporary commentary. I feel there is value in admitting my mistakes and acknowledging the dead ends I pursued. In doing so, I hope I assist you and save you from spending time exploring the same avenues.

Some additional works have been included. There are two articles from 1999, first published on my uidesign.net blog. There is a guest article by project management expert Mike Griffiths. There are also several articles I wrote for other forums, including my paper for the proceedings of the Agile 2005 conference.

Many people reading this text will be new to Kanban; they will not have read my earlier book. Some of the contemporary commentary may be tantalizing for those people. I choose not to fully explain my Kanban-related comments. For a fuller explanation, you will need to look elsewhere. I hope you become sufficiently curious to learn more. My book, *Kanban: Successful Evolutionary Change for Your Technology Business*, is a good place to start.

For those who are already familiar with Kanban, I believe the selected works presented here will fill in gaps in your understanding and provide an explanation as to how Kanban came about and where its origins truly lie. I hope this depth of understanding will enable you to be a better Kanban practitioner and to provide deeper, more insightful guidance to your peers, your staff, your clients, and your audience (whether readers of your own blog, or attendees at conferences where you are presenting).

There were a number of articles that we love, and that over the years have been among the most popular on the site, but they didn't always fit into the flow of the narrative and the arrangement of chapters. However, we didn't want to omit them. These articles were often light-hearted and humorous, and we felt that their inclusion would enhance your enjoyment as you invest your time digesting this volume. We've inserted 16 of these lighter pieces between the chapters.

Yet some of the best articles, such as Superstition and Boiling Frogs, didn't make the cut. Inevitably, some fans of Agile Management blog may be disappointed that their favorites haven't been included. On the other hand, I hope you will appreciate our editorial choices and find this book a useful addition to your library. The lessons in Agile management included here are truly the ideas, concepts, and stories that shaped my current work and many of these lessons are actively included in my classes. I am sure my clients reading this will recognize much of the material. What may surprise them is how old it is!

I hope you will enjoy this collection of my work from the last 13 years. I encourage you to join the Agile Management group on Yahoo! to ask questions and provide your own thoughts.

David J. Anderson
Sequim, WA, April 2012

Happy Groundhog Day
Thursday, February 2, 2006

TODAY IS GROUNDHOG DAY.[1] PERHAPS NOW IT'S BETTER known for the movie[2] of the same name. Here is my ironic Groundhog day resolution . . .

I will continue to do what I've always done, while expecting things to improve and the outcome to change!
I will continue to do what I've always done, while expecting things to improve and the outcome to change!
I will continue to do what I've always done, while expecting things to improve and the outcome to change!
I will continue to do what I've always done, while expecting things to improve and the outcome to change!
I will continue to do what I've always done, while expecting things to improve and the outcome to change!
I will continue to do what I've always done, while expecting things to improve and the outcome to change!
I will continue to do what I've always done, while expecting things to improve and the outcome to change!
I will continue to do what I've always done, while expecting things to improve and the outcome to change!
I will continue to do what I've always done, while expecting things to improve and the outcome to change!
I will continue to do what I've always done, while expecting things to improve and the outcome to change!
I will continue to do what I've always done, while expecting things to improve and the outcome to change!

1. http://www.groundhog.org/info/feb2.shtml
2. http://www.imdb.com/title/tt0107048/

I will continue to do what I've always done, while expecting things to improve and the outcome to change!

I will continue to do what I've always done, while expecting things to improve and the outcome to change!

I will continue to do what I've always done, while expecting things to improve and the outcome to change!

I will continue to do what I've always done, while expecting things to improve and the outcome to change!

I will continue to do what I've always done, while expecting things to improve and the outcome to change!

I will continue to do what I've always done, while expecting things to improve and the outcome to change!

I will continue to do what I've always done, while expecting things to improve and the outcome to change!

I will continue to do what I've always done, while expecting things to improve and the outcome to change!

I will continue to do what I've always done, while expecting things to improve and the outcome to change!

I will continue to do what I've always done, while expecting things to improve and the outcome to change!

I will continue to do what I've always done, while expecting things to improve and the outcome to change!

I will continue to do what I've always done, while expecting things to improve and the outcome to change!

I will continue to do what I've always done, while expecting things to improve and the outcome to change!

I will continue to do what I've always done, while expecting things to improve and the outcome to change!

I will continue to do what I've always done, while expecting things to improve and the outcome to change!

I will continue to do what I've always done, while expecting things to improve and the outcome to change! . . .

On Leadership

K ANBAN REVEALS SYSTEMIC ISSUES AND CREATES THE OPPORTUNITY FOR CHANGE. THIS CAN BE SCARY, ESPECIALLY IN AN ORGANIZATION THAT HAS BEEN RESISTANT TO CHANGE. THERE IS A DESIRE FOR PERMISSION. AN ACT OF LEADERSHIP CAN BE THE SPARK THAT IGNITES THE NEEDED CHANGE. It is usually a manager or a team lead who first embraces change by signaling that a specific change is needed and that it can safely happen. Leadership can start a cycle of improvement that can lead to a major cultural shift.

Leadership also reflects values. We see it in the decisions managers make and in how the workers are treated. Lean encourages us to take a systemic approach to performing work and to view workers both as a part of the system and as stakeholders in its effective operation. The goal is to finish the work—and the system should support that goal in a healthy and balanced way. It is an imbalanced system that drives the workers harder and harder.

Strong leaders must make bold, risky moves that might be unpopular but ultimately are necessary for success. We can judge a decision retrospectively by looking at the outcome, but also by the congruence of the decision with the values of the organization and the integrity of the individual. The Agile movement itself has produced a number of leaders who shake things up—some for good, others not so much.

One of the founders of the Agile movement, Jim Highsmith's first book was called *Adaptive Software Development* (Dorset House, 1999). Its core idea, "inspect and adapt," became a core value of the Agile community. Kanban has taken this idea further by suggesting that we make guided changes based on an understanding of workflow models. We experiment with small changes, keep them if they work, and discard them if they don't. We call this an evolutionary approach to change. But evolutionary changes need a catalyst to get them started. It all begins with an act of leadership.

New Rules for Old Geeks

Friday, January 19, 2007

• •

The challenge for us as managers is to give [the profession of] management a good name by putting in place Lean processes that facilitate rather than hinder, that deliver both productivity and work–life balance, that lead to great code and healthy coders, and that continue to delight customers without all-night code merges and death-march schedules.

• •

THIS WINTER I'M CELEBRATING 25 YEARS IN THE SOFTWARE INDUSTRY. I'M ALSO facing the arrival of a mid-life crisis as my fortieth birthday approaches this spring. Yes, it is 25 years since a group of 14 year old schoolboys (the linked article dates from 1985) launched an advertising campaign in the classified ads in the back of Sinclair User magazine advertising games for the Sinclair ZX81. In exchange for a check in the princely amount of £3.50 we would mail you a set of listings of games written in BASIC. You had to type them in to your computer in order to play! :-O

Back in those days motivating geeks to write great software was easy—just feed them pizza and cola and let them work all hours of the night and don't worry about all that homework that isn't being done.

The conventional management wisdom is that the software industry is different. Software programming geeks are different. Motivating them is different. You don't manage them. You herd them. The idea goes like this: You hire smart people, usually as young as possible and with as little social life or social skills as possible, stick them in an open-plan office, let them decorate their cubes any way they like, bring toys in to the office, provide a ping-pong table, a foosball table, a fully stocked kitchen, free juice, and a budget to order in food after hours, and then just leave them alone. The result will be fantastic innovation and don't worry about the quality, the bugs can always be fixed in a future version. This wisdom has prevailed ever since, and here on the west coast of the USA it's a formula that has made many executives and venture capitalists rich, so they would see little reason to change a winning formula.

But have you noticed anything recently? Perhaps when you are sitting in a meeting providing status on your latest project? Grey hair maybe? Balding heads? Bifocals?

As a manager are you noticing that staff need a lot more time out of the office for medical appointments and other real life events? engagement? marriage? birth of child? death of parent? illness? injury? kids stuck home on a snow day?

When you're recruiting, have you noticed how carefully the applicants read the benefits package literature and negotiate for flexible spending plans and childcare facilities, and how disinterested they have become in the location of the ping-pong table? Have you been asked whether prostate screening is covered under the medical plan?

The '80s young-gun geeks are still here. They are still producing great software. They still love their jobs and love the profession. They take a pride in what they produce. BUT . . .

They have kids to get home to. If a project runs into trouble, they'll still pull an all-nighter and brag that they've still got what it takes as they feel a proud burst of nostalgia for the old days, but three days later they'll be out sick struck down by the latest flu variant and you'll lose a week of productivity as a result.

The industry is aging!

We need new management rules for old geeks (like me). These rules mean establishing processes to ensure a good work–life balance. Old geeks want their capability to produce balanced against the demand from the business. They won't be death marched. They already regret missing out on their 20s. The gallus[1] geek of yesteryear who talked disdainfully of process, carried his compiler in a holster slung low from his hips, and treated management as the pointy-haired boss to be ridiculed, now sees process as his friend and his boss as the facilitator of balancing professional success against the demands of real life.

The challenge for us managers is to give management a good name by putting in place lean processes that facilitate rather than hinder, that deliver productivity and work–life balance, that lead to great code and healthy coders, and that continue to delight customers without the all-night code merges and death-march schedules.

Old geeks need new rules. Old geeks are great software engineers; they still have a lot to offer. The successful organizations of today will learn how to provide a well managed environment that delivers on the needs and wants of the middle-aged, graying developer population. The others will continue to play by the old rules and will suffer continual churn and turnover as they burn out an increasingly intolerant workforce.

---

1. From Firstfoot.com, a Scottish vernacular dictionary: ADJ. Bold, daring, rash, wild, unmanageable, impish, mischievous, cheeky. As in "Yi gallus wee besom, eh'll skelp yir erse!"

Prairie Chickens

Wednesday, March 10, 2004

•••••••••••••••••••••••••••••••••••••

"Because failure was punished and success could be achieved by sitting back and watching the money pour in, the executives who made it to the top were those who avoided mistakes, not those who made bold moves." —Bob Lewis

•••••••••••••••••••••••••••••••••••••

I'VE BEEN GIVING SOME DEEP THOUGHT TO THE CONGRUENCE OF OUR IMMEDIATE goals and alignment with shareholders interests. To help understand the problem, I would like to introduce you to the prairie chicken. This is not my work but that of Bob Lewis who writes the Survival Guide column for *InfoWorld*. Bob's 1999 book, *IS Survival Guide: Changing CIO from "Career is Over" to Change is Outstanding* (Pearson Sams Publishing, 1999) is one of the best books on management—not just for software development but for any manager in an internal IT department. The following is an excerpt from page 15, which I quote in full.

Prairie Chickens and Old-School Executives

Modern American business executives are expected to visualize, promote, and lead change. This is profoundly different than the experience of American business even twenty years ago, when executives had a lot in common with the male prairie chicken. Let me explain.

Prairie Chickens are less elaborate cousins of the peacock, living in places like the plains of western Minnesota and the Dakotas. Every spring in the hour or so around dawn, male prairie chickens congregate in areas about the size of an average lawn called "leks," claim personal territories, and do the prairie chicken dance. It's an amazing sight.

The central territory is the smallest—about three feet across. Further from the center territories get bigger but less desirable, as least as defined by female prairie chickens, which wander through the lek choosing which males will be allowed to fertilize their eggs. Proof: The central male gets about 90% of the matings.

For as many years as biologists were aware of the prairie chicken dance, they assumed the central male was the toughest, nastiest bird for miles around. At least, they did until a graduate buddy of mine named Henry MacDermott looked into the matter. Henry discovered something completely unexpected. Male prairie chickens don't fight hard to carve out their territories, defending them against all comers. As the years pass, males sort of drift to the middle. There's some competition, but for the most part the males who survive long enough—those that don't die of disease or being eaten by foxes—end up near the middle.

When I left graduate school to enter the world of business in 1980, I often found myself wondering how some of the executives I encountered managed to get to their positions of influence and authority. Most of them treated important decisions the way you and I would treat a rabid ferret: something to be watched carefully and avoided while someone else foolish enough to stick his or her neck out deals with it.

One day I figured it out. In the early 1980s, American business was suffering from thirty years of complacency, succeeding more from momentum built up in the '50s and '60s than from current quality. Japan Inc. was kicking American business in the shorts, but our executives had acquired their positions the previous decade or two.

Executives who took risks that failed were punished for their failures, and businesses succeeded nicely without taking any risks at all. Because failure was punished and success could be achieved by sitting back and watching the money pour in, the executives who made it to the top were those who avoided mistakes, not those who made bold moves.

American executives, in other words, were prairie chickens!

I personally feel that this is not just an "old" problem. There are plenty of young executives who were inflated in rank during the bubble years in high technology and telecommunications who would qualify as fully fledged prairie chickens. How many do you know? Perhaps you work for one?

Failure to Manage to the Values

Tuesday, July 31, 2007

• •

Knowledge workers' productivity is directly related to
their motivation and engagement in their work. . . .
Hence, I came to the conclusion that living by the
values is paramount.

• •

FOLLOWING ON THE THEME OF GOALS AND ALIGNMENT, I'D LIKE TO CONSIDER
a management dilemma posed by Jack Welch, former CEO of GE. It's modeled as
a classic 2 × 2 matrix (or Gartner Magic Quadrant chart), as shown in Figure 1.1.

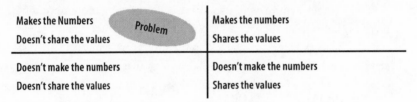

Makes the Numbers Doesn't share the values	*Problem*	Makes the numbers Shares the values
Doesn't make the numbers Doesn't share the values		Doesn't make the numbers Shares the values

Figure 1.1 Goals and values chart

On one axis is the notion that a manager represents the values of the organization, and the other axis shows whether the manager delivers results (the
numbers). This gives us four possible outcomes. Welsh states that in three of the
outcomes the management decision is easy:

- ❑ makes the numbers, shares the values—no brainer—this manager is
  excellent and a keeper

- ❑ doesn't make the numbers, doesn't share the values—no brainer—fire
  him or her

- ❑ doesn't make the numbers, shares the values—a good manager, well
  intentioned, needs an investment in coaching in order to improve the
  numbers

However, it's the fourth type that causes the problem. What do you do with
someone who makes their numbers but does so in a way that is out of alignment
with the values of the organization? In the short term, eliminating someone like

that might mean that the numbers are not achieved and that the higher level manager takes a blow to his or her own numbers. How do you explain a J-curve effect period that results from actions undertaken to fix the values of the organization in a world where values are about managing for the long term, but where the markets expect us to manage for the next quarter?

In software projects, I see these managers as those who deliberately, consciously death march their people to achieve a project date, or who deliberately, consciously sacrifice quality for the same reason, or who manage in some other fashion that might deliver results but leaves a sour taste—for example, a manager who suffers 50% staff turnover, rules with an iron fist, and has instilled the fear of death in to the 50% who do hang around long enough to get anything done.

It's all too easy to see managers revert to methods that are not in alignment with Agile values, particularly when there is a specific extrinsic incentive for them (and usually not for the workforce.) The war of attrition attitude—deliver the numbers, let me make my bonus, and don't worry about the casualties! It's a style that shows that the manger is being narrowly managed and doesn't have to pay the costs of attrition—that comes from a different budget. The system isn't geared to inflict the penalty on the offender. System effects like this can be seen in general life, for example, large industries get to pollute our environment and aren't always taxed suitably to cover clean up or management costs..

My view on this is very simply—knowledge workers are human. You cannot deny their humanity. To invoke Jeff De Luca's First Law, "[It's] 80% psychology, only 20% technology!" Knowledge workers' productivity is directly related to their motivation and engagement in their work. All the process and methods and organization in the world will not fix a demotivated workforce. And hence, I come to the conclusion that living by the values is paramount. You cannot sacrifice the values to hit the numbers in the short term. So for me the Welch dilemma isn't a dilemma at all. Managers who don't live by the values should be removed. Period!

Question: Does your organization have published values? Do your workers understand the values you use to lead your organization?

Lead People, Manage Things!

Thursday, August 5, 2004

••••••••••••••••••••••••••••••••••••

Although there may be one right way to manage
projects, there is no right way to manage the people
who are engineering the project—those people must
be led.

••••••••••••••••••••••••••••••••••••

THE AGILE COMMUNITY HAS, OVER THESE LAST THREE YEARS, CONTINUED TO
debate self-organization* and whether there is a need for any management at all
in software development. Peter Drucker offers us this advice in *Management
Challenges for the 21st Century* (Harper Business, 1999):

> One does not **manage** people.
> The task is to lead people.
> And the goal is to make productive the specific strengths and
> knowledge of each individual. (Emphasis original)

In my book, *Agile Management for Software Engineering* (Prentice Hall,
2004) and again in an article in the current edition of *Cutter IT Journal* ("The
Four Roles of Agile Management," July 2004), I argue for splitting project management (the oversight of the delivery to the customer) and engineering management (the oversight of the system of software engineering) into two distinct roles
filled by different people. Why?

Drucker wrote the above words in the context of an old assumption: that
there is one right way to manage people (or there should be). He argues that
this view is inaccurate and that although there may be one right way to manage things, people cannot be managed by prescription. We could interpret this
for software engineering by saying that although there may be one right way
to manage projects, there is no right way to manage the people engineering
the project—those people must be led. In simple terms, the project manager is
responsible for managing things—the issues and day-to-day problems that arise
in a project—by removing the obstacles and facilitating the flow of work. The
engineering manager has a different job—to lead the people, to organize a system
for software design and construction that motivates them to do good work as a

team and to build that team into a learning organization that is always looking for ways to do better work and become more productive. It's a structure with the engineering manager as leader, project manager as manager of things to be delivered, with the team reporting to the leader rather than the manager.

● ● ● ● ● ● ● ● ● ● ● ● ● ● ● ● ● ● ●

* This debate has continued now for a decade and it is still a popular topic on discussion lists and at open-space events.

To Code or Not to Code?

Wednesday, January 21, 2004

• •

Don't let subjective debate and lack of belief cost you
inertia. Lead by example.

• •

RECENTLY I'VE BEEN CODING!

Back in the days when I worked at IBM as a contract coder—in the mid 1990s—my line managers were young and had recently been promoted to their first management job. Management training involved as many as five weeks away from the team and was referred to as "the brainwashing" by their former colleagues. One of the key tenets drilled into these new managers was the idea that they must never again get directly involved in the day-to-day development work.

The position of line manager is key in any firm. The line manager is the one who manages the economic engine of the company. Without good line management, there is no business. The transition from individual contributor to line manager is also perhaps the most difficult and the most important for any business. For the first time, new line managers must learn to live vicariously through their staff. They must learn to coach the best performance out of their team and to accept that some jobs will never be done as well as they would like or as they themselves might do. This behavior of accepting the job as a vicarious pursuit and learning to settle for "good enough" rather than "perfect" is a measure of how well a manager is progressing in his or her new role—and something on which further promotions to higher levels might depend.

It seems obvious, then, that if former developers are to learn to live vicariously, they must never code again. The temptation to "just do it" might otherwise be too great and they will never become good managers—always down in the noise rather than defining the governing rules for their systems and the metrics with which to monitor and control its capability.

Management is all very well; however, occasionally the troops need guidance as well as leadership. Developers need to be shown the way. This is particularly likely in new and uncertain territory. The greater the uncertainty, the greater the leadership required. Uncertainty can be caused by the domain—a new market being entered, perhaps—or the technology, or the tools being used, or the working practices being adopted. In any or all of these circumstances, if the manager has hands-on experience, then it makes sense to lead by example.

When introducing Feature-Driven Development, for example, don't be afraid to lead a feature team as chief programmer! Don't be afraid to demonstrate Design-by-Feature by convening a feature team and facilitating the drawing of a Sequence Diagram. Don't be afraid to hold a code review. When developers are unsure of how to implement behaviors such as "blue" <<description>> classes on a domain model, don't be afraid to pair-program with a developer to make it happen. Pair programming is perfectly acceptable in Feature-Driven Development as part of a project's early lifecycle while the architecture and design templates are being created. Don't be afraid to fuzzy the line between step 1, Build a Domain Model, and step 4, Design by Feature. It doesn't matter if you have to build a dozen features to prove a model. If "seeing is believing" then make your development team believe. Lead them to believe through your example. Don't let "analysis paralysis" bog you down. Don't let subjective debate and lack of belief cost you inertia. Lead by example.

It is okay for managers to code when they are leading! Once momentum builds and the team feels comfortable, then the manager can quietly slip back into the vicarious life of monitoring metrics, studying capability, and considering the policies that define the system in operation.

Compensate for Uncertainty
with Leadership

Sunday, February 1, 2004

● ●

Without sufficient leadership, there will be the chaos that Satir identifies; but rather than recovering from the slump to show improvement—the J-Curve effect—there is simply a slump and things remain worse than they were before.

● ●

THERE HAS BEEN CONSIDERABLE DEBATE IN THE AGILE COMMUNITY ABOUT THE percentage of good people required to make Agile methods work. The topic is also a common running theme in Boehm and Turner's book, *Balancing Agility and Discipline* (Addison-Wesley Professional, 2003), where, on page 185 (for example), they postulate that FDD requires a lot of good people in order to scale. Schwaber and Beedle's *Agile Software Development with Scrum* (Prentice Hall, 2001) claims on page 121 that for Scrum to work successfully, at least 50% of the people need to be experienced. My own experience with FDD is that one senior, experienced developer for every six team members is sufficient to scale the FDD method to a large project. Why?

To be candid, I believe that this community endeavor to define how many good or experienced people are required for any given method is a red herring. The real issue is one of leadership versus uncertainty. Uncertainty comes in many forms—change; ambiguity in scope, domain, or job description and position; schedule uncertainty; fear; difficulty; novelty of process, technique, or tools, and lack of experience therein; and the consideration for assignable-cause variations. The greater the uncertainty in any project, the more leadership is needed (at all levels) in order to compensate for it. By loading too much uncertainty into a project without sufficient leadership, there will be the chaos that Satir identifies[1]; but rather than recovering from the slump to show an improvement—the J-Curve effect—there is simply a slump and things remain at a level worse than they were before.

---

1. Satir, Virginia, et. al., *The Satir Model: Family Therapy and Beyond,* (Science and Behavior Books, 1991)

Feature-Driven Development gets its leadership at many levels: The first is Chief Programmer (the shop-floor foreman), then the Development Manager, and finally the Chief Architect. All these roles must provide leadership on a day-to-day basis in order to overcome the uncertainty. When a project is happening in a new domain, with uncertain scope and schedule, and FDD is being introduced for the first time—along with color modeling and perhaps new middleware and development—then everything is in flux. In such a situation, a team needs every Chief Programmer, Development Manager, and Chief Architect to be great leaders, mentors and teachers. In simpler situations, where there is less ambiguity, less leadership is required.

In order to be a leader, an individual must have the team's respect. For technical people, this generally means they must be both good at software development and experienced in the process in use and in the full software lifecycle from inception of a project to delivery.

The amount of leadership required on a project is situational and contextual. In my opinion, trying to measure it by method and define a scale of Agile methods by the percentage leadership required is futile!

Management by Reality—
Fired on the Asphalt

Thursday, January 6, 2005

•••••••••••••••••••••••••••••••••••••••

The differences in style between Trump and Branson
are real . . . I have made up my mind who I'd prefer
to work for. Have you?

•••••••••••••••••••••••••••••••••••••••

THIS PAST FALL SEASON HAS SEEN A TV RATINGS WAR BETWEEN THE KING OF
Manhattan real estate, Donald Trump, and the mildly eccentric (superficially)
Sir Richard Branson. Trump had a second season of his show, *The Apprentice*,
while Branson launched his copycat show, *Rebel Billionaire*.

The most interesting thing to watch was the very obvious differences in
styles. Trump seems to breed confrontation, and his organization is very hi-
erarchical—a hierarchy reinforced with the trappings of power. The final two
contestants got to boss around former colleagues while being chauffeured in
a Maybach limousine as their lackeys followed behind in a minivan. Trump
dismisses the losers with his punch line, "You're Fired!" The entire season was
filled with bickering, Machiavellian intrigue, and outright bitchiness among the
colleagues—particularly the women.

Meanwhile, Branson's style is collaborative. He allows the losing team to debate who should be up for his "elimination challenge." Branson encourages consensus, while Trump encourages finger pointing and dissent. Whereas Trump
looks for loyalty to the defeated leader, Branson looks for objectivity in analysis
of the defeat. Branson then has the losing elimination pair challenge each other.
In the case of an outright loser then he doesn't have to fire anyone—they self-
selected, such as the guy who fell asleep during the night while camping out in
the African savannah. An emotional Branson hugged the guy as he handed him
his ticket home on the airport canopy before they boarded their next flight. He
was evidently sorry to lose such a strong candidate.

The final dismissal with Branson is again non-confrontational—the un-
charitable might call it passive aggressive. He simply confronts the two losers
on the tarmac at the steps to the plane and hands them each a ticket. One ticket

allows someone to board while the other sends the loser back home to the USA on a different flight.

One final key difference is that Branson never asks his elimination contestants to do anything he wouldn't do himself, and he often joins them. With Branson it's all casual clothing, breaking bread around the table, hugs, and emotional support. It wouldn't be a stretch to imagine that his organization is much flatter than Trump's and that his senior managers don't enjoy the trappings of power because he recognizes that the only power they wield is the power to influence through respect.

When I'm watching *The Apprentice,* I can't help but remind myself that it is entertainment and that much of it is set up for the viewer. It isn't real. It isn't reality. It's fake! However, I feel that the differences in style between Trump and Branson are real and really are reflected in the nature of the two game shows. I have made up my mind who I'd prefer to work for. Have you?*

* * * * * * * * * * * * * * * * *

* It is worth considering that Branson's show ran for only one season, while *The Apprentice* ran for five and spawned other versions around the world. The conflict, the politics, and the power and its trappings appealed to the viewing public much more than Branson's collegial collaborative culture did.

Some Military Lessons

Sunday, January 30, 2005

• •

A team can fracture into a group of individuals because trust breaks down when mistakes are made.

• •

FROM TIME TO TIME, I LIKE TO HIGHLIGHT EXAMPLES OF BAD LEADERSHIP, THIS time from the most recent episode of *The Apprentice* (now in the third season). In this episode, Verna walked out. Yip! She packed her bags and walked off early on the second morning. Although Michael was the project manager, it required Carolyn to show what a true leader would do under these circumstances. She got in her car and went to look for Verna.

I couldn't help feeling that last season's winner, Kelly, a former military man, would have known what to do—you never leave anyone behind! Where was Michael after Verna disappeared?

Something that led to Verna's departure was her lack of sleep. In fact, both teams seemed to pass on sleeping at all in order to maximize their input. The leaders were managing hours of effort rather than effectiveness. Something else military people understand is that some sleep is essential. When a platoon sleeps, two soldiers keep watch. After an hour or two, they wake colleagues and swap. In a critical situation, more stay awake, but the platoon commander always ensures that every team member gets a couple of hours of sleep. We seldom, if ever, see this on *The Apprentice*.*

And finally, a theme that comes up again and again throughout the three seasons—food. Every good military manager knows that an army marches on its stomach. It's a key part of management and leadership to ensure that workers are well fed. If a manager asks staff to work overtime, then the manager shows up and makes the coffee, fetches the pizza, or does whatever it takes to keep people productive. Again and again on this show, we see project managers neglect to ensure that their team gets fed.

These are such basic mistakes. There are three golden rules we can learn from this glaring example of poor management practices.

No one gets left behind. It's very bad for morale. When the team sees that management will abandon a weak member, it makes them feel insecure. Deming said that we should "drive out fear."[1] You can't eliminate fear if you show that you'll leave someone behind.

Everyone gets some sleep. People start to make mistakes very quickly when they are tired. Mistakes cause trouble and agitate others. It becomes a vicious cycle as tired people irritate each other more and more. More mistakes get made. More time is wasted. Focus on the task is lost.

Everyone gets to eat sufficiently. Without food people don't think straight or act rationally. Mistakes get made (see above). This irritates people and undermines trust among team members. They see others at their worst and they react to it. A team can easily fracture into a group of individuals because trust breaks down when mistakes are made. Sleeping and eating are at the foundation of Maslow's hierarchy of needs;[2] don't let the lack of either of them cause the team to fall apart.

● ● ● ● ● ● ● ● ● ● ● ● ● ● ● ● ● ●

* In 2006, I had the pleasure of meeting George H. Ross, Donald Trump's lawyer and a star on *The Apprentice*. In a short but truly valuable conversation that I cherish to this day, he told me that the show's editing process was partial to showing mistakes. When leaders were taking good care of their people, it wasn't interesting for the audience, so it was cut. Hence, we only ever saw the bad examples and seldom, if ever, the good. As a result, the show appears to have educated us on bad management and poor leadership. A sad reflection on what makes for good viewing numbers.

1. Point 8 in W. Edwards Deming's "Fourteen Points of Quality" (http://deming.org/index.cfm?content=66)

2. http://www.abraham-maslow.com/m_motivation/Hierarchy_of_Needs.asp

Staff Ride

Wednesday, November 9, 2005

• •

Just because the military expects soldiers to follow
orders, it doesn't mean that they aren't expected
to think for themselves and to act locally based on
immediate feedback.

• •

FOUR YEARS AGO TONIGHT, I WAS SLEEPING OUT UNDER NINETEENTH-CENTURY
canvas with the rest of the Sprintpcs.com leadership team, in the cold, damp,
gloom of the western Arkansas hills. Why on earth? We were on a Staff Ride!

A Staff Ride is leadership training normally given to US military officers.
The "ride" refers to touring a battlefield on horseback. The "staff" are a leadership group. Nowadays internal combustion powers the "ride"; no horses were
involved. The battlefields we toured and the battles we relived were both from
the American Civil War—the Battle of Wilson's Creek[1] and the Battle of Pea
Ridge.[2] Each member of the ride gets to role-play an officer from one of the two
sides of the battle. These were chosen at random the evening before. Some of us
were generals. I played lower-ranking leaders, including a battery commander.

The Civil War was in its own way an information-age war. The battle records written up by each officer after combat serve to this day as a record of who
did what and when. It's possible to stand on the exact site and pretend to aim
the cannon at the enemy marching over the opposite hillside. Remarkable after
140 years! The battlefields are preserved by the federal government—lest we all
forget. An appropriate sentiment as we approach November 11—Armistice Day,[3]
as we Brits call it.

The staff ride training was the brainchild of Chris Tabor. Chris had served
in the US Marine Corps in both the Gulf War and later in Somalia. If you've
seen Black Hawk Down,[4] you will understand the role Marines played (or didn't)

1. http://www.nps.gov/wicr/

2. http://www.nps.gov/peri/

3. http://en.wikipedia.org/wiki/Armistice_Day

4. http://www.imdb.com/title/tt0265086/

in that battle. In more pleasant times Chris served in London. He worked for my boss in a staff role similar to the military position of adjutant. Chris ran our monthly operations review about which I wrote chapter 14[1] of *Agile Management for Software Engineering*. The ride was organized by Military Professional Resources Inc. (MPRI),[2] and was conducted by three colonels who also had served in the first Gulf War.

If ever I had any doubt that military officers "get" management and leadership, it ended with this outing in the hills. This was by far the best training event I've ever participated in. I learned such a lot. For example, you've heard the term "death march"; perhaps you've even read Ed Yourdon's book about it. But have you ever seen one—truly? Can you imagine the general who marched his troops around the back of the mountain in the middle of the night in the dead of winter and then expected them to attack the fresh and rested enemy from the flank first thing after dawn? When you stand on the site where the soldiers fell, you learn the meaning of death march. It's not pleasant.

I also learned how military command and control works—how to separate out strategic intent from operational missions and tactical battlefield events. I learned how the delegation rules work and how the rules of engagement are written based on whether something is tactical, operational, or strategic: "Lieutenant, we need to eliminate the gun emplacement at the top of this hill so that we can move our support forces through the valley below. Take your platoon and eliminate it for us. Signal me when you have control of the location." What the lieutenant does after that is tactical and under his control. Operational-level rules might involve commands such as, "General, march your division through Missouri and take control of the Kansas river; engage the enemy on discovery and eliminate them." Finally, a strategic-level decision might involve rules like, "We want to deny supplies of food and other matériel to the enemy. Supplies are transported on the Mississippi–Missouri river. Therefore, we must take control of the river and prevent the enemy from using it." This three-level separation of strategy, operations, and tactics are the three basic layers of control and delegation for any corporation. The top sets the strategy, the middle executes on the operational plan, and the lowest levels act tactically in response to local conditions.

Do not confuse military leadership with Henri Fayol[3]–style command and control within the industrial setting. They are not the same. Just because the military expects soldiers to follow orders, it doesn't mean that they aren't

1. http://www.agilemanagement.net/index.php/blog/Sample_Chapter

2. http://www.mpri.com/

3. http://www.analytictech.com/mb021/fayol.htm

expected to think for themselves and to act locally based on immediate feed-back. As managers seeking to lead agile teams, we can learn a lot from military officers.

There is a lot more I could write about these three days and two nights in Arkansas in 2001, but my memory fades. The picture shows the commemorative ship's compass that was given to each of us as a memento of the occasion. It sits on my desk to this day. The inscription reads:

"Leadership is the art of accomplishing more than the science of management says is possible," —Colin Powell.

Handy on Failure Tolerance

Monday, January 15, 2007

• •

If you are trying to be lean and eliminate waste but
you're not driving out fear and encouraging learning
from failure then you won't succeed.

• •

I'VE BEEN READING CHARLES HANDY's *Beyond Certainty*[1]. IT'S A COLLECTION OF essays and speeches produced over many years. They read like blog entries. Each one is relatively short, and during this difficult winter period when I'm commuting on the bus rather than biking to work, I've been enjoying several chapters on each trip.

In a piece entitled "Are there bugs in our offices?" where Handy is discussing the need to move to leaner, flatter organizational structures, he remarks on lack of trust as a virus that is infecting our offices and breeding inefficiency. He goes on to point out that failure tolerance is the key to enabling a leaner, flatter organization . . .

> Leaner, flatter management structures only work if they result in more junior people having some senior responsibilities. The savings come in fewer controllers and requests for permission, fewer inspections and inspectors. More responsibility, however, means that more people will act on their own initiative, and that inevitably means more mistakes. Punish those mistakes, inscribe them in the corporate memory, and you will make it quite certain that no one will exercise their initiative again. The layers of command and checking will build up once more; the savings will vanish.
>
> If the leaner, flatter structures are going to work we have to invest a lot of effort in helping those in the front line to make the right decisions, not in punishing them if they make the wrong ones. *Asking for help when in doubt must be seen by everyone as a sign of responsibility, not a symptom of weakness.* Mistakes if they occur can be a wonderful way to

---

1. Handy, Charles, *Beyond Certainty: The Changing Worlds of Organizations*. Cambridge, MA: Harvard Business Press, 1996.

learn, perhaps the only way to learn, as long as we are prepared to admit that they were mistakes and do not try to defend or excuse ourselves. Fear makes all this impossible; fear locks the organization into rigidity, making it conform to yesterday's rules which may not be the right rules for today's problems.

I chose to highlight the sentence that calls out a key behavior that we encourage in agile teams and agile organizations. However, this whole piece really drives home the point that lean, flat structures, high trust–low waste organizations that tolerate failure and drive out fear tend to have all these attributes. If you are trying to build a high-trust culture but you've got 14 layers of hierarchy, there is a mismatch. If you are trying to drive out fear but you are not failure-tolerant, there is a mismatch. If you are trying to be lean and eliminate waste but you're not driving out fear and encouraging learning from failure, you won't succeed. If you aren't encouraging people to ask for help as a demonstration of their responsibility to the shared goals of the team, then you are letting fear fester and you are discouraging the less confident from contributing fully to your productivity and success.

Leadership Heroes #1: Alex Ferguson

Monday, May 7, 2007

• •

Ferguson has never shied away from an unpopular decision if he felt it was right for the team's long-term performance.

• •

So it seems that Manchester United won the English League soccer title[1] over the weekend. It's timely, then, that I'd planned my second post on heroes (my first actual one on managers) to be about Alex Ferguson.[2] (Unlike my fellow Singapore project[3] colleague Stephen Palmer,[4] I'm not a United fan but I am a huge admirer of their coach.)

Back in 1986–87 I was working as a games developer for Ocean Software and their joint venture, U.S. Gold. Ocean was based in Manchester, England. I mostly worked from home, flying or driving to Manchester perhaps one week per month. Alex Ferguson had just taken over as the manager of Manchester United, following his short spell as manager of the Scottish international team that played in the 1986 World Cup Finals. Ferguson was already an accomplished manager. He'd taken unfashionable Aberdeen to glory in European tournaments and won the Scottish League and other titles, as well as managing the national team. In some respects he didn't have much else to prove.

He was still living in Glasgow and commuting to Manchester by air, presumably going home after the game on a Saturday and flying back on a Monday or Tuesday for team training and the following week's game. The staff of British Airways treated him like royalty. He was invisible. They'd keep a seat in the front row of the BAe1-11 jet (Americans can think MD-80) and "Fergie" would enter the plane and take his seat just before they closed the doors to push back and taxi.

---

1. http://news.bbc.co.uk/sport2/hi/football/teams/m/man_utd/6630511.stm

2. http://en.wikipedia.org/wiki/Alex_Ferguson

3. http://en.wikipedia.org/wiki/Feature_Driven_Development

4. http://www.step-10.com/

So why is Ferguson one of my management heroes? Well, first off, there is longevity. In 21 years he has had only one job and he's been hugely successful at it. He's had this one job in a field that is not known for longevity in management careers. This is partly a tribute to Ferguson himself and partly a tribute to his hiring manager—Martin Edwards, the Chairman of Manchester United at the time. Longevity and loyalty in management careers both by an individual and their management are to be admired.

The next aspect of Ferguson's career is again a tribute to the patience of Martin Edwards.* After more than three years without winning a trophy, the fans wanted a new face on the coach's bench. But Edwards stuck by Ferguson, telling people they had the right man for the job. And they did.

Another reason I admire Ferguson is that he manages prima donnas all the time. The staff on his payroll earn huge salaries—millions—at a very young age. Ferguson has the mental toughness to deal with this and to keep his players under control and focused.

Ferguson is also known for his success at bringing up young players and grooming them for stardom and future places on the regular first team. By developing their football academy and through it bringing up young star players, Ferguson paid nothing to acquire talent like David Beckham, Paul Scholes, Gary and Phil Neville, and Nicky Butt. A true manager knows how to grow and nurture talent, and Ferguson has repeatedly shown this ability over the years.

But the single most important reason I admire Ferguson is his ability to focus and make tough decisions about team selection and players. Often a player can be at the height of his fame and be hugely popular with the fans, but Ferguson will see in training that the player is past his best: perhaps running slower, perhaps diminished fitness or strength or declining skill and accuracy because of it, or other issues off the field of play. Ferguson has never shied away from an unpopular decision if he felt it was right for the team's long-term performance. He's stayed focused on keeping Manchester United a winning team, and after 21 years he's still doing it.

So, management hero #1, Sir Alex Ferguson.

● ● ● ● ● ● ● ● ● ● ● ● ● ● ● ● ●

* At the time of writing, Sir Alex recently celebrated his twenty-fifth anniversary in his job at Manchester United. This prompted a flurry of international press coverage, including biographies in papers such as the International Herald Tribune—not a paper that generally reports the careers of soccer team managers. Interviewed on the BBC, Martin Edwards revealed more of the detail of his and

the board's decision to stick with Ferguson through the 1989 season and beyond. The board contained Bobby Charlton, former United player and World Cup winner with England in 1966, and Sir Matt Busby, the former manager of that team, also from Scotland, for whom Ferguson was the spiritual successor. They had hand-picked Ferguson for the job and they believed that he was always the right man with the right experience. It's evident now that they believed in him absolutely. It may also be true that if they asked the question, "If not Ferguson, then who?" there were few, if any, names on their list.

Emotional Elizabeth

Saturday, January 29, 2005

My wife inspired today's blog entry, while we were watching an episode of *The Apprentice* last season. Elizabeth was having yet another crying scene, and my wife blurted out, "Ahhh, Elizabeth is just such a woman!" I almost choked on my fruit tea. So, Elizabeth was showing her emotional, sensitive side—I guess these are attributes more commonly found in women. Of all the people I know, perhaps only my homemaker wife can get away with a comment like this. It's certainly not something an American manager can express explicitly in the workplace; and it made me think. Although I don't see such sentiment expressed explicitly, I do hear managers complain about behavior in men and women that could be described as merely "human." Just as in the show, with Elizabeth, these expressions of humanity are often looked upon negatively—as though they were a failing. So, to pick on this specific example—do we really want all women in the knowledge workplace to be hard, unemotional, and thick skinned? Do we want them to be able to take everything in their stride—all business, totally focused? Personally, I find this idea a bit frightening.

People come in a whole range of varieties, and managers have to understand that. Some don't cope well with change, or uncertainty, or lack of direction. People feel pressure in different ways—they worry about things that often aren't worth worrying about. It's the manager's job to understand human behavior and to assign roles and responsibilities appropriately. Furthermore, it's the manager's responsibility to control chaos, to bring stability to an organization, and to act rationally in the face of variation—to understand what is normal variation and what represents exceptional circumstances that require intervention. Stability helps staff to function better, regardless of their emotional or cultural disposition. In the event that one of the team does have a breakdown, it's the manager's job to fix it.

The implication that someone "can't handle it," and there is "no room on the team for someone who can't pull him- or herself together" is really a manager treating the symptom—taking the easy way out. It's easy to say, "I don't want this person around." The harder job is to look for the root cause and deal with that; to take away the circumstances that caused the emotional breakdown so that it doesn't happen again. A lack of leadership is what failed Elizabeth, and the nature of this game show, which has one person eliminated each week, didn't help. Elizabeth was identified as a weak link and quickly became an outcast—someone who could be sacrificed for the survival of the others. This made the situation worse. It became a vicious cycle as she felt more and more isolated.

This is where a show like *The Apprentice* helps us all—by highlighting examples of poor leadership.

On Management

I N *SOFTWARE ENGINEERING ECONOMICS* (PRENTICE HALL, 1981), BARRY BOEHM CONCLUDES, "POOR MANAGEMENT CAN INCREASE SOFTWARE COSTS MORE RAPIDLY THAN ANY OTHER FACTOR." THE LESSON IS THAT THE HIGHEST LEVERAGE FACTOR FOR IMPROVING THE EFFECTIVENESS OF SOFTWARE development is in educating managers to manage better.

The emphasis on the system in Lean thinking leads directly to a discussion of management. Managers are rightly responsible for the system—both its design and operation. A system is made of people performing activities, all of which are governed, regulated, and coordinated by policies. Managers make policies and they have the power to change or override those policies. Hence, the effectiveness of the system results directly from management's judgment and decision making. The manager faces a path laid with hazards and opportunities. Hazards exist where managers can poorly influence workers—through inappropriate measurement, destructive policies, or simple inaction when exceptional circumstances interrupt the normal running of the system. Opportunities exist in the form of personnel development, permission giving, and encouraging learning.

In Lean theory, variability is a factor that we watch closely. Deming[1] said that performance would improve only if the organization could learn to understand and reduce its variation. Management policies, such as a strong "definition of done," can guide the choices workers make to reduce undesirable variability, while managers can design a system that is robust enough to cope with desirable variety, allowing late breaking-changes and work of different types and risks to be treated differently.

1. http://deming.org/index.cfm?content=66

The Agile world has had a love-hate relationship with management. It has encouraged an end to the heavy paperwork burden of olden times. It has empowered developers and kept managers out of the development timebox. While encouraging the definition of boundaries of empowerment, there has been little guidance on how managers can add real value. Nor has there been much impetus from the Agile community to address Boehm's core finding from thirty years ago—that improving managers would have the greatest economic impact on software development. On the other hand, management support is almost always required for an Agile change initiative to succeed.

We must acknowledge the importance of management and the valuable role that managers can play in improving agility and economic outcomes from software development activities.

Manager as Permission Giver
Thursday, January 22, 2004

• •

A manager can act as this tipping person to change
both the behavior of his team and the performance
of his organization. A manager can change the
culture simply by giving permission to change it.

• •

EVER SINCE I READ THE *TIPPING POINT* BY MALCOLM GLADWELL (LITTLE,
Brown, and Co., 2000), I've been toying with the concept of the role of manager
as permission giver.

On pages 223 through 230 Gladwell talks about "Tipping People," or permission givers. He uses negative, destructive behaviors such as suicide and teenage smoking as examples, but the concept can be applied positively, too. People
can tip the balance and cause a cascade of behavioral change by either leading
by example or giving permission. A manager can act as this tipping person to
change both the behavior of his team and the performance of his organization.
A manager can change the culture simply by giving permission to change it.
Here are some examples:

MANAGER: Quality is poor! You should start code reviews before
check-in.

DEVELOPER: But we have no time to do code reviews!

MANAGER: I give you permission to take the time to do code reviews.
Please start them.

MANAGER: Scaling this project is difficult. It takes too long to bring
new people up to speed. You need to produce design documents.

DEVELOPER: But we have no time to do designs!

MANAGER: I give you permission to take the time to do designs.

MANAGER: The finished code does not match the design documents.
You should do design and code reviews.

DEVELOPER: But we have no time to do reviews and keep documentation in sync!

MANAGER: I will get you a tool that keeps the code and design in sync. I give you permission to take the time to learn how to use the tool and to use it to the fullest of its capabilities. I further give you permission to take the time to perform reviews to ensure that the code is in sync with the design.

...AND SO ON

• • • • • • • • • • • • • • • • •

* In 2006, I used this approach with my team at Corbis. I gave them permission to stop creating bugs. Before I joined the firm as the senior director for software engineering, each release had, on average, generated six hot fixes due to critical defects escaping into the wild. Within six months, there were no defects, critical or otherwise, escaping. In 2007, 42 releases of software produced only 21 defects and none of them were critical. This was achieved simply by creating the cultural norm that it was okay to take the time to finish something properly, to a high standard, and with confidence that it was working properly. Given a choice between making a deadline or taking the time to complete work with high quality, developers had my permission to make the second choice. If anyone, particularly a project manager, was unhappy with this choice, then that person was referred to my office. It's amazing how powerful the effect was of knowing that senior management supported the individual knowledge worker who chose to do the right thing.

Management Misdirection

Monday, September 27, 2004

• •

Developers are right to be wary of managers who ask for or require individual measurement. It is primarily a misdirection tactic that the manager can use to avoid blame.

• •

IS YOUR MANAGER A MAGICIAN? SOMEONE WHO THROUGH SLEIGHT OF HAND can misdirect attention away from what is really going on? Can he or she always avoid blame and slide through his or her career unhindered by a failure to deliver? It seems this is all too common. I have been recognizing a number of attributes that contribute to the survival ability of bad managers but now I want to focus on just one—individual measurement as misdirection.

I want to put a new slant on why measuring individuals is bad. Bad managers ask for individual measurements because they want to know who to blame; it's a way to identify a scapegoat and separate out the weakest link. Bad managers do not see it as their own failing that they don't develop their people. They expect individuals to "learn on their own time," or to "take responsibility for their own careers." Naturally, there is a need for individual knowledge workers to sustain their own knowledge—it is, after all, the tool of their trade. However, managers have that responsibility, too. Better individual performance adds to better team performance.

Developers are right to be wary of managers who ask for or require individual measurement. It is primarily a misdirection tactic that managers use to avoid blame. When things go wrong, the weakest performer is singled out to take the fall. This assumes that the manager in turn works for an equally weak middle-manager who accepts excuses, finger pointing, and sacrificing of the weakest by means of the measurements gathered. Perhaps it is how they themselves got promoted?

How to Destroy Morale: Lesson 1

Tuesday, September 28, 2004

• •

Trading off quality in exchange for fast hacking is setting up a development team for failure. It's classic management misdirection.

• •

"Soooo, Bob, how's it going?" said Morris. If only he wore suspenders, it could have been a scene straight from the movie Office Space.

"I know you and the team have been working hard this year. You've got a great architecture, development is running smoothly, and you've made the date for all three iterations. The customer is happy, too. The early releases have been passing all the tests and our customer has really learned to trust us. However, there is just one thing . . ."

"I'm going to need you to," pausing, and then slowing his speech, "forget about quality and for the rest of this year, just hack it out. Forget quality, we need speed!"

Silence.

Bob reaches down and picks up his chin with his right hand, physically lifting it to close his mouth again. After a short pause, he mumbles ,"Errrr, okay. We'll see what we can do." The irony is that Morris had, earlier that year, stood in front of the executives and said, "The bottom line is that quality makes us go faster!" Everyone had smiled and applauded his efforts with Agile software development.

So what is wrong with this scenario? Well, yes, Morris is a rotten manager, but that's not really what I'm getting at. Morris has destroyed his staff's morale in just two sentences. But there is more. What's wrong in Morris's organization is that they are measuring the wrong thing. At the executive level, they are measuring "code complete." It is the code complete date that gets reported on the wider program level. Meanwhile, the marketing guys are asking for too much and failing to recognize the value of thorough analysis and the truth in objective velocity data. You can't put a quart of anything into a pint pot. So what to do? "Hey, just hack it, and as long as we're only measured on code complete, then we're home free," thinks Morris.

Morris may be a natural talent at management misdirection. He successfully sets up the developers for failure. If they miss the date, they fail; then he can claim they were "too academic" and "perfectionists." If they ignore quality and by magic hit the date but then they spend months fixing bugs, they fail and they take the blame for the lousy quality. After all, the manager can't possibly be to blame for such unprofessional conduct. Perhaps we can't blame Morris. He's just a survivor in a bad organization. He isn't measured on deployed working code and his senior management are unresponsive to objective data. They always believe that developers can be squeezed to produce more—knowledge work is a soft target; brains are spongy, aren't they?!

Trading off quality for reduced lead time is a false economy. It will bite you. Trading off quality in exchange for fast hacking is setting up a development team for failure. It's classic management misdirection. Magical misdirection isn't good up-management, it's employee abuse. If your manager is an accomplished magician, it might be time to polish up your resume.

The Line Manager Squeeze

Monday, January 3, 2005

• •

The line manager squeeze—setting an impossible task for the line manager—is another example of management misdirection. Senior management conveniently divert attention and deflect blame and criticism onto the junior managers.

• •

SOMETIMES, OBJECTIVITY, TRANSPARENCY, AND EFFECTIVE UP-MANAGEMENT have no effect. Why? Because senior management chooses to squeeze the line management temporarily to compensate for their own failings.

Recently, a friend of mine quit his second job within 12 months. Why? Because, "I'm sick of being set up for failure," and "It was the same old b***s*** over the schedule." In short, he was a manager being squeezed.

While I was back in Scotland in October, I had dinner with an old college buddy and his wife, who had just quit her airline job as a number-one flight attendant—the cabin crew's line management position. Why? Because she was being asked to sacrifice quality of service (that is, her and her staff's pride in their jobs) and staff break times in order to sell more merchandise. In short, her employer's business model was broken: They couldn't make enough money selling airline tickets, so they tried to plug the gap by spending more and more staff time pushing goods onto a captive market. The ever-increasing retail sales targets meant that service and staff breaks had to be sacrificed on shorter flights, or the targets would be missed. A line manager wishing to keep her job had to make herself unpopular with her staff and her customers. Of course, the longer-term effect is to damage the brand and the airline's reputation—but what the heck, the executive management will have retired by then; and the line managers will still be doing the same old job, trapped by personal commitments such as home loans and kids to put through college, unable to break free and take the financial risk of quitting to look for better, more fulfilling opportunities.

The line manager squeeze—setting an impossible task for a line manager—is another example of management misdirection. Senior management conveniently diverts attention and deflects blame and criticism onto junior managers. It's yet another example of why the line management job isn't an enviable one. No wonder so many senior techies prefer to stick to technical positions as systems architects or senior developers and don't volunteer for leadership training and management positions.

Why Aren't Managers Paid More?

Tuesday April 10, 2007

• •

I believe that we must embrace Barry Boehm's observation from 1981 *Poor management can increase software costs more than any other factor.* So far, we're an industry in denial of this basic truth.

• •

IN *THINKING FOR LIVING* (HARVARD BUSINESS REVIEW PRESS, 2005), THOMAS H. Davenport argues that managers of knowledge workers should be paid a premium for letting go of the relative safety and security of their individual-contributor knowledge worker jobs and assuming management positions. His reasoning is simple: Managing and organizing knowledge workers is vital to their productivity; self-organization and empowerment only go so far, then management has to step in. However, a first-level management job requires an individual to take a huge personal risk, to abandon the skill set that made him or her successful and learn a whole new skill set as a manager. In order to attract the appropriate candidates into management, goes Davenport's thinking, it is necessary to pay a premium. How much of a premium is hard to say, but ten to twenty percent seems appropriate. If this sum aggregated for the remainder of the individual's career, then the risks might seem worthwhile.

Interestingly, compensation professionals tell me that the market does not support premium salaries for software-engineering managers. I think that there is one main reason for this, and that a number of secondary reasons might indicate a root cause of the problem. The first reason is basic economics. Due to uncertainty in the nature of the work, managers must maintain a flexible workforce of contingent labor (contractors): Typically, ten to fifty percent of his or her personnel are contingent, hourly paid, temporary staff. The contingent nature of contract labor requires that a premium be paid to compensate the worker for the lack of security and continuity of employment. Good people, confident in their technical skills and their ability to renew and refresh those skills regularly, can earn a premium as contractors, often earning more than middle managers and junior executives. Put another way, a geek can earn more as a contractor than he or she could make suffering through ten years of climbing the corporate ladder as a manager. Hence, permanent, full-time, individual-contributor knowledge worker jobs also fetch higher rates. Given these factors, there is no premium for managers. **In fact, the market would suggest that managers should really be paid less!**

Clearly this is a problem. If effective management is the highest leverage factor in knowledge worker productivity, as Barry Boehm observes in *Software Engineering Economics* (Addison-Wesley Professional, 2003), there is a conflict when the open market does not remunerate managers appropriately. What is causing this conflict?

I think there are a number of causes. First, I feel that geeks always look for a technical solution to a problem before they pursue a people or process solution—in other words, tools over operational innovation or sociological/psychological influence within a workforce. If the answer is always to deploy a new software-based tool, the demand will always be for high-end geeks who can make the best software. Secondly, technical innovation and problem solving is valued over operational innovation and the ability to manage to a plan or to use higher-maturity quantitative and probabilistic approaches to management. The high-tech tribe tends to value individual ability to create great product innovation and solve significant problems in computer science rather than process innovation and the soft skills it takes for a manager to motivate a team.

In conclusion, I feel that if we are to deliver on Davenport's vision—that good management will come from offering a premium for knowledge workers to make the leap to a new skill set—we must first start to value management skills more highly. In order to value management skills more highly, I believe that we must embrace Barry Boehm's observation from 1981: *Poor management can increase software costs more than any other factor.* So far, we as an industry have been in denial of this basic truth. Until we face, as Collins and Porras suggested, "the brutal facts of our own reality"[1]—that pursuing a management career negatively affects the earning potential of top tech talent—there is little hope for fixing the situation.

---

1. *Built to Last: Successful Habits of Visionary Companies* (HarperBusiness, 1994)

Why Good Managers Still Matter
to Agile Development

Tuesday September 9, 2003

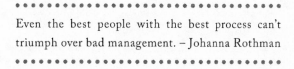

Even the best people with the best process can't triumph over bad management. – Johanna Rothman

Johanna Rothman writes in her blog[1] that managers still matter, even with Agile development processes.

> But what about managers? Earlier, I said that managers matter too. Here's why. Good people can triumph over inadequate process or inadequate management. But even the best people with the best process can't triumph over bad management. Bad management trumps everything else. I've worked for bad managers, and I bet you have too, so you've seen the damage bad managers can cause.

This topic is close to my heart. It has a lot to do with why I wrote *Agile Management for Software Engineering* (Prentice Hall, 2004). Bad managers kill software productivity. I recall a conversation I had with the then-CIO of Sprint PCS, Jerry Batt. We both agreed that management in IT was universally poor on the average. I said I thought it was because not enough managers had been programmers. Often, people get into a PM track too early in their careers and as a result, developers don't respect them. He replied that he believed the reason that managers were so poor was that too many of them were programmers. We held the same view and agreed on its effect, but we disagreed on its cause.

I now realize that I was talking about leadership, while he (as a trained senior manager) was talking about management—goal setting, decision making, measurement, analysis of feedback, actions, interventions, and investment strategy.

1. http://www.jrothman.com/weblog/archive/2003_08_01_mpdarchive.html#106207585754525870

Policies: You've Got the Power!

Friday June 15, 2007

• •

If a policy is affecting the performance of your team
—change it! Show some courage. Make a change.
Just do it!

• •

IF YOU ARE A MANAGER, CHANCES ARE THAT YOU ARE RUNNING A DEPARTMENT
with sub-optimal performance. Much of that performance is being constrained
by policies. Those policies might have been around for years. Many of them predate your taking control of the team. Many of the staff have forgotten why any
one policy was introduced. Those policies are part of the folklore of how things
are done around here. Probably your shop-floor individual contributors know
that these policies are affecting performance but they abdicate any responsibility,
viewing it as a management problem.

Guess what? You are the manager!

Policies are under your control (or maybe your boss's or a senior up-line
manager's).

If a policy is affecting your team's performance—change it! Show some
courage. Make a change. Just do it!

For example, our department had a policy that locked down our test environment for three days after every release. Since introducing our kanban system
for sustaining work, we were putting a release out every two weeks. A three-day
outage in testing was constraining our productivity by limiting our test capacity
to only 70 percent efficiency. The effect was lower total throughput of change requests and delivered business value. A small cross-functional team investigated
the history behind the three-day lockdown policy and discovered that most of
the original reasons for it no longer applied. On the team's recommendation,
I changed the policy to one day, which gave us back 20 percent of our testing
capacity. Since we made this change in early April, we've seen continued growth
in throughput of processed change requests delivered to production each month.

Where are your policy-constrained bottlenecks?

Flipping Metrics

Friday April 22, 2005

• •

We should be able to show that we are learning
and that this learning manifests itself as reduced
variation. Under no circumstances should we use
this data to evaluate individuals' performance.

• •

No I'm not swearing.

This is a serious post about the difference between management by objectives (or conformance to plan commitments) versus management of process variation. The management scientists will recognize this as the Peter Drucker versus W. Edwards Deming debate (Drucker and Deming actually roomed together at one point in the early '50s, when this argument started. Drucker is said by Deming to have capitulated eventually.) Yet others might see this as a Crosby[1] versus Deming debate. And in many ways the Crosby idea of measuring quality by conformance to specs isn't so far removed from the Drucker idea of setting a target, getting a commitment from an accountable person, and then holding their feet to the fire until they deliver.*

Back in 2001, Mac Felsing and Ken Ritchie were working with me at Sprintpcs.com. I was managing a dev team, in addition to other duties, and they were working for the newly appointed Senior Director of Engineering in his process improvement group. They were tasked with building adoption for Feature-Driven Development (FDD) across the business unit. One day they arrived at a meeting with me filled with frustration. The boss had asked them to start measuring individual performance on feature commitments and report conformance against commitment.

I'm on record as being very anti–individual measurement. I believe that measuring individual feature milestone commitments and using them in performance reviews would blow an FDD team apart and destroy both productivity and morale. It's a fundamental tenet of FDD that you never, ever, measure individuals. Heck, FDD doesn't even monitor individual tasks—only the features

---

1. Philip Crosby, first known for his concept of DIRFT (doing it right the first time), from his book, *Quality Is Free* (McGraw Hill, 1979).

constructed by a team. I felt I had to save the senior director from his own ineptitude. So I suggested that they flip this metric on its head and take it back to him as a report on our ability to estimate—a measure of the variation between planned and actual. Over time, we should, as a team and an organization, be able to show that we are learning and that this learning manifests as reduced variation. Under no circumstances should we use this data to evaluate an individual's performance.

The executive request was for management by objectives (MBO) and conformance to plan; any deviation would have individuals paying the consequences. Management knew that layoffs were coming the following year and they wanted to identify the weaker players with objective data. The alternative to MBO is management of process variation. (I promised myself that I would get out from under this new boss; five months later I quit Sprint altogether.)

Deming firmly believed that individuals were almost never responsible for their own performance, but rather victims of the variation in the tasks that they had to perform. This is often referred to as the 95/5 rule—that only five percent of the outcome is affected by individual performance. Now there is a radical idea—programmer productivity is not responsible for the performance of the team or organization! It seems that results like Sackman's, from the 1960s, although they might show large variance in individual performance, could mean that this variation doesn't make a huge difference in the overall outcome. **

● ● ● ● ● ● ● ● ● ● ● ● ● ● ● ●

* This is perhaps the most radical idea in technology management alive and kicking inside the Kanban community today.

● ● ● ● ● ● ● ● ● ● ● ● ● ● ● ●

** Given that Drucker empirically observed and correctly interpreted many complex aspects of knowledge work, it is ironic that he failed to spot the negative effects of setting targets.

Gardeners

Saturday April 23, 2005

• •

The "can't do the job" employee is not responsible for his or her performance. The manager is responsible.

• •

I'VE BEEN CHALLENGED BY MY LONGTIME COLLEAGUE AND LOYAL RIGHT-HAND man, Daniel Vacanti, to explain what to do about underperformers. This is a perennial issue: What do you do with team members that you, and everyone else, knows is underperforming?

I think there are two types of underperformers—those who simply can't do the job, and those who won't do the job. I'd like to deal with the first type here.

"There are a lot of gardeners in IT" my former boss, Jeff De Luca used to say.

This expression dates from his years with IBM and the grim, dark years under the leadership of John Akers. The world economy was reeling from the 1987 stock market bust and the subsequent property bubble, and IBM was laying off people left, right, and center. In discussions with his boss at the time, Jeff had been assured, "There is always work for good people" and, "There are a lot of gardeners in IT" who might soon be "mowing lawns around Melbourne" once again.

Dan will recall the time we interviewed a tree doctor for an $80-per-hour contingent staffing position as a Java developer. His previous IT experience amounted to less than three years as a contractor doing Java development at a large wireless company and several years in the Cascade mountains healing sick trees. Following the interview, during which he couldn't describe the difference between the == operator and the .equals() method, I advised his agent to recommend that he go back to his true profession. In general, I try to avoid the gardener problem by not hiring them in the first place.

The employee who can't do the job is the responsibility of the manager. It is the manager's failing. The employee is still trying to do his or her best—even when it isn't nearly good enough.

As Dan rightly points out, a gardener on the team can destroy morale. When you inherit a gardener from a previous manager, you've got a problem. You identify gardeners primarily through a "managing by walking about" strategy. For me this includes the sub-classes "managing by having lunch," "managing by drinking coffee," and "managing by playing ping-pong or foosball." Many of my former staff will recognize these techniques. Once identified, the next step in dealing with a gardener is to stop including his or her "productivity" in your estimates. Once the team has effectively benched the gardener, you as a manager can't count him or her as part of the productive team when estimating projects. By eliminating the gardener from the estimate, you prevent the team from carrying the burden through extra heroic effort. That just leaves the morale problem. The gardener has to be managed out. As anyone working in the USA's Fortune 500 knows, this is a tricky problem and it takes time—sometimes up to 18 months.

I've mentioned before that I like to measure individuals on secondary contributions. For example, "Become the language lawyer on unit testing and test-driven development and transfer your new knowledge to the other team members. Build their respect as the authority on TDD and make yourself the go-to point for advice on TDD and unit tests." Now, that is something I can measure. A gardener will never be able to accomplish that task. Managing the gardener out is the responsibility of the manager. Coaching the team quietly and privately to be patient and to understand the difficulty of achieving this is how to deal with the morale problem.

In summary, the "can't do the job" employee is not responsible for his or her performance. The manager is responsible. The manager never should have hired him or her in the first place.

Attitudinals

Sunday April 24, 2005

• •

They can choose to work functionally with the team or they can strike a pose and behave like spoiled toddlers. But when it comes to writing code, they aren't responsible for their performance.

• •

THIS IS PART 2 OF MY REPLY TO DAN. FOLLOWING YESTERDAY'S POST ABOUT gardeners, here I discuss those who simply won't do the job—those with a posture or attitude that makes them dysfunctional on the team.

It's easy to argue that the attitudinal is responsible for his or her own performance. He can choose to change his attitude and start to perform. Were this to happen, his performance might well come up to or exceed that of the average team member. As a team, we'd certainly be able to factor his performance into our estimates. We could take him off the bench.

Team members with an attitude problem usually are not difficult to spot. They simply don't behave according to established team norms. For example, their attendance at the morning meeting is sporadic and when they do attend, they don't act like functional team members. The body language (and sometimes the spoken language) is all wrong. It's also possible to have a passive-aggressive, fifth columnist[1] who hides his dysfunction and expresses it with poor quality code or very slow code production. This person is harder to spot and requires intelligence from the grapevine—best gathered through the "managing by having coffee (or lunch)" approach.

So once again poor performance is the manager's responsibility. The manager must talk to the individual with the attitude and make it plain why and how he needs to change. Then the manager must not tolerate more bad behavior or bend the rules. There can be no exceptions. It's like parenting.

I've seen all sorts of attitudes. The most common, perhaps, is "I'm God's gift to software engineering," often coupled with the complaint, "The team doesn't give me any of the cool stuff to work on." My advice: Try doing the mundane

---

1. The fifth column originally referred to Franco sympathizers during the Spanish Civil War; now it generally refers to someone who acts subversively.

stuff and doing a good job of it; earn the trust and respect of your peers. If you do the mundane stuff well and they learn to trust you, they will eventually realize that you can help them to be successful and they will give you some of the more difficult, sexier tasks.

Again, it's important to take the attitudinal out of the estimate. Don't let the rest of the team carry the burden. Treat the person as bench slack or give him individual tasks to perform that are off the deliverable's critical path. There is a danger here. A side effect may be that—the attitudinal is isolated from the team and can never be integrated. If your goal is to turn this individual into a fully functioning member of the team then you must put them in a position and give them work that will earn the respect of the wider team.

By maintaining a firm stance on expected norms, the attitudinal will either (a) change his attitude, earn the trust and respect of other team members, and reintegrate as a functional team player, or (b) self-select out of the organization.

So, to Dan's point, I believe that individuals are responsible for their professionalism and their discipline, and for their attitude toward their job, the business, and their fellow team members. They can choose to work functionally with the team or they can strike a pose and behave like spoiled toddlers. But when it comes to writing code, they aren't responsible for their performance. That will always vary according to the techniques employed. Improving the techniques, reducing the variation, and increasing the mid-point (or mean) capability level are the responsibility of the whole team and the manager. Team members simply have to agree to be a part of a learning organization—to strive to be professional and to show a willingness to learn. By learning to treat dysfunction as a separate management problem, we free ourselves to focus on improving capability.

How to Communicate with Me!

Tuesday January 17, 2006

· ·

It's important to develop a good reputation for up-management—to show bosses that you can handle the work on your own and only escalate matters when you truly need a more senior manager to help.

· ·

IT'S IMPORTANT TO DEVELOP A GOOD REPUTATION FOR UP-MANAGEMENT—TO show bosses that you can handle the work on your own and escalate matters only when you truly need a more senior manager to help. It's also important to show that you can take an issue to the right level for resolution: Don't bother a GM when a Senior Director will do; don't bother an SVP when a plain old VP will do. But that's a topic for another post.

I developed the following template while I worked for John Yuzdepski. He understood that service goes downward in management and he encouraged us to let him know how he could be of service to us. I use the template I developed for communicating with John as a way to train my staff to better up-manage. It's important not to expect people to do this intuitively. Generally, their only up-management training came when, as children, they learned how to manipulate their parents to get what they wanted. Manipulation isn't the result we're after. Understanding the correct level at which to make decisions and how to ask for senior intervention is what we are looking for. Here is the template:

To: John
From: David
Subject: An issue I'd like your help with

Background
A paragraph (or two) explaining the background. Focus on information the executive doesn't already know.

Issue (or Problem)
Describe the precise issue or problem. There might be a web of linked issues or a series of causes and effects that needs description, but try to keep it brief and to the point.

Proposal (or enumerated options)

Describe how you'd like to solve the problem, or a set of alternatives. Describe the pros and cons of the proposal or set of options.

Action (or, "What I'd like you to do for me today.")

Describe precisely how you would like the executive to help you. What action do you want her to take, or what decision are you asking him for? Make the request actionable, or explicitly state that you need the executive to escalate the matter and help identify which level is appropriate.

When I first hand out this template at a team meeting, it is amazing to watch the team members' faces. They don't quite believe it is real. Then I wait for the first one brave enough to actually try it; it might take weeks, but eventually someone does. I can see the relief on the person's face when they realize that it actually works. Gradually the word spreads. Using a formal template makes the sender think harder. It makes them realize that this is an official memo. It also cuts down on over-communication and noise in your inbox. Too often team members will copy you on every email—stuff you don't need to read. Sometimes, something important will get missed. Seize such an opportunity to educate people that they have to tell you when they want your help—you're not psychic! Introduce the template. Noise in your inbox will drop off.

Warning: Don't try this on bosses who fundamentally don't get the idea of servant leadership. One guy I worked for would yell, "Stop telling me what to do!" You have been warned.

The Defective Paper Towel

Tuesday, January 10, 2006

"BRING ME A PAPER TOWEL, WOULD YOU?"

"Certainly!" I walk to the kitchen and rip a paper towel from the roll behind the sink. I deliver it to its destination. As I approach, I realize the purpose for said towel. A small child is pushing to escape the bindings of her high chair, her face pasted with tomato soup and fragments of broccoli and pasta woven into her hair.

"I guess you'd like it wet?"

"Yes, I want to wipe her face."

"You didn't specify that in the requirements. I'll need to charge you extra!"

"Charge me extra?"

"Yes, your requirements were faulty. I've incurred extra work."

"Charge me extra! You should know that requirements always change."

It's that simple!

What Would Drucker Do?

P ETER DRUCKER IS PROBABLY THE KING OF TWENTIETH—CENTURY MANAGEMENT SCIENCE. HE COINED THE TERM "KNOWLEDGE WORKER" AND WENT ON TO DEFINE IT. HE SURELY HAD PROFESSIONS SUCH AS SOFTWARE DEVELOPER IN MIND WHEN HE ENVISAGED ENTIRE INDUSTRIES FILLED with knowledge workers. He later acknowledged that learning how to manage, motivate, and lead knowledge workers would probably be the greatest management challenge of the twenty-first century. Over the years, I have found inspiration in Drucker's writing and I often have been surprised by his insights from as early as the late 1950s and early 1960s.

I believe the Agile movement would have resonated with Drucker, were he still alive to appreciate it. Themes like avoiding big up-front planning, focusing on effectiveness rather than on efficiency or utilization, and the fundamental concepts of empowerment and avoidance of command-and-control management were also core themes for Drucker. He inherently understood that in knowledge work, by definition, workers understand how to do the work better than their managers do, and that the traditional supervisory role in which managers direct the work is unlikely to succeed.

Peter Drucker told us to lead people and manage things. With Kanban, I've encouraged a focus on the work by means of a visual display such as a kanban board. The progress of each work item can be observed as it moves toward completion. This lets us see how the work flows (or doesn't) through the steps required to complete it. We manage the work by observing and influencing flow and by defining the rules of the system through which it flows. Workers are empowered within the boundaries of those rules and are free to complete their own tasks as they see fit. The emphasis is on efficiency of flow rather than efficiency of work, and their utilization of idle time creates pressure for improvement and both the freedom and the time to work on process improvement. By encouraging team members to feel motivated and positive about their contribution, to be focused on the main goals of the business, and to build a learning organization, we create an environment that promotes productivity. I believe Drucker would have recognized this as a major step in the right direction.

The Next Great Challenge

Originally posted as "Drucker Month" on August 2, 2004

● ●

Constant excess of demand is one of two root causes
of management challenges in the technology sector.
The other is that the work is invisible. Invisible work
is impossible to manage.

● ●

PETER DRUCKER IS CONSIDERED THE FATHER OF MODERN MANAGEMENT
science. He started studying and writing about management in the 1940s; he
was the academic who made it fashionable. In his early years he worked (or at
least shared an office) with W. Edwards Deming. Later Deming and Drucker
had some professional disagreements, primarily over Drucker's "Management
By Objectives" (MBO), which Deming deeply disagreed with. We'll be looking at
Deming's work in Chapter 4. Meanwhile, to kick things off, here is an observation from Drucker's 1992 book, *Managing for the Future* (Dutton, 1992):

> Managing the knowledge worker for productivity is the next great chal-
> lenge for American management.

Writing in the same year, Ed Yourdon predicted "the decline and fall of
the American programmer" (in a book so named), as he felt sure programming
jobs would be outsourced offshore to lower-cost economies. I recall giving a
recruitment presentation for interns in 1994 at Napier University in Edinburgh
and predicting, "Programming is a dead-end profession when your job as a programmer can be done in India for one-fifth the cost." I was offering to give the
interns exposure to "more than just programming" and an "opportunity to learn
skills that will protect your career." Of course, both Yourdon and I were wrong.
What saved the industry was the Internet bubble—demand exceeded supply and
cost was no object of concern until the bubble burst. Here in 2004, we're learning it wasn't a cure, but a stay of execution, and once again knowledge worker
productivity is the next great challenge for the American manager.

It seems that history repeats itself. While the technology sector continues
to grow in India, China, Brazil, and much of South America and Asia, layoffs
in the rich West are still few and far between. Most knowledge workers quickly
find new employment. Ceaseless innovation and experimentation means that

there is no shortage of ideas for new software, and demand continues to exceed supply. I've become convinced that this constant excess of demand is one of two root causes of management challenges in the technology sector. The other is that the work is invisible. Invisible work is impossible to manage. Invisible work is easy to stockpile and hoard. Together, excessive demand and the invisible nature of the work result in continual overburdening. This is why I believe Kanban is proving so effective (but that is a topic for another book).

So what does Peter Drucker have to say about productivity of knowledge workers and what might we in the software engineering profession learn from it?

Drucker on Effectiveness

Monday August 9, 2004

• •

The assumption is that local efficiency leads to global efficiency—a local optimum hoping for a global optima. In comparison, modern Lean/Agile thinking is about effectiveness of a whole system.

• •

A KEY DIFFERENCE BETWEEN AGILE DEVELOPMENT METHODS AND THEIR PREDECESSORS is a focus on effectiveness rather than on utilization, which is the focus of the cost-accounting definition of efficiency. Others in the community have picked up on this theme, too. The relationship between efficiency and cost accounting goes back to Frederick W. Taylor and his time 'n' motion studies from the turn of the twentieth century. "Taylorism," as it is known, is about specialization and efficiency (utilization) of specialists—the assumption is that local efficiency leads to global efficiency—a local optimum hoping for a global optima. In comparison, modern Lean/Agile thinking is about effectiveness of a whole system. It uses a systems thinking approach. To achieve this it uses more adaptable, generalist workers rather than specialists optimized for utilization.

Peter Drucker has had quite a bit to say about effectiveness and knowledge workers. Most of what follows is taken from *The Effective Executive* (Harper and Row, 1967); note that it is a clear four years before *The Psychology of Computer Programming* (Van Nostrand Reinhold, 1971) by Gerald Weinberg. In the mid-sixties, software engineering had something to learn from general management science but at the time the link had not yet been made.

To be effective is the job of the knowledge worker.

. . .[P]eople of high effectiveness are conspicuous by their absence in knowledge jobs. High intelligence is common enough amongst knowledge workers. Imagination is far from rare. The level of knowledge tends to be high. But there seems to be little correlation between a man's effectiveness and his intelligence, his imagination, or his knowledge. Brilliant men are often strikingly ineffectual; they fail to realize that the brilliant insight is not by itself achievement.

Intelligence, imagination, and knowledge are essential resources, but only effectiveness converts them into results.[1]

---

1. Drucker, Peter F. *Managing for the Future* (Dutton, 1992)

Drucker then separates the Taylor era from the modern knowledge worker era so effectively with this definition:

> For manual work, we need only efficiency, that is, the ability to do things right rather than the ability to get the rights things done.

Efficiency = Do Things Right
Effectiveness = Do The Right Things

> Increasing effectiveness may well be the only area where we can hope significantly to raise the level of the knowledge worker's performance, achievement, and satisfaction.

> One of the weaknesses of young, highly educated people today . . . is that they are satisfied to be versed in one narrow specialty and affect contempt for the other areas.[1]

Drucker comments that finding effective people is difficult and observes, "[T]here is no 'effective personality'"; today we would summarize that as, "There is no Myers-Briggs Type for 'Effective.'"

He goes on to observe that effectiveness is not achieved from principles, but from successful execution of practices. If we are to learn from Drucker, we cannot merely talk about Agile development as abstract principles—even though this is the simplest way to gain consensus in the community. In order to show effectiveness, we must articulate practices that deliver productivity:

> Effectiveness is a habit; that is, a complex of practices. And practices can be learned.

Hence, Drucker leaves us with hope—hope that if we teach, coach, mentor, and instill the correct set of practices in individual knowledge workers, effectiveness will follow. What is largely lacking in teaching software engineering as a profession is just that—a set of practices that lead to effectiveness. If we can fix that both at the academic level and within the industry itself, we can start to deliver the value that shareholders deserve from their expensive knowledge workers.

---

1. Drucker, Peter F. *Managing for the Future* (Dutton, 1992)

Drucker on Refactoring

Tuesday August 10, 2004

Over-zealous refactoring destroys shareholder value.
So beware of the cruft polisher on your team.

DRUCKER ON REFACTORING—NO, REALLY! WHAT FOLLOWS IS WHAT PETER Drucker might have had to say about refactoring were it available as a practice when he wrote these words in 1954:

> A favorite story at management meetings is that of the three stonecutters who were asked what they were doing. The first replied, "I am making a living." The second kept on hammering while he said, "**I am doing the best job of stonecutting in the entire country.**" The third one looked up with a visionary gleam in his eyes and said, "I am building a cathedral."
>
> The third man is, of course, the true "manager." The first man knows what he wants to get out of the work and manages to do so. He is likely to give a "fair day's work for a fair day's pay.
>
> It is the second man who is a problem. Workmanship is essential; without it no business can flourish; in fact, an organization becomes demoralized if it does not demand of its members the most scrupulous workmanship they are capable of. But there is always a danger that the true workman, the true professional, will believe that he is accomplishing something when in effect he is just polishing stones or collecting footnotes. Workmanship must be encouraged in the business enterprise. But it must always be related to the needs of the whole.
>
> . . . The tendency to make the craft or function an end in itself [in future] will therefore be even more marked than it is today. . . . The new technology will need both the drive for excellence in workmanship and the consistent direction of managers at all levels toward the common goal.[1]

What Drucker is telling us is that craft workmanship such as zealous refactoring is important both to an organization's morale—it is important that people can take pride in their work—but also to the quality of what is being produced.

---

1. *The Practice of Management* (New York: Harper & Brothers, 1954)

However, over-zealous refactoring destroys shareholder value. So beware of the cruft polisher on your team. Refactoring should be done for the right reasons—to eliminate technical debt, to rebalance the books, or to facilitate future iterations. It should never be done simply because a developer doesn't like the architecture or the implementation. If you can't answer, "Yes" to the question, "Does this code prevent us from efficacious delivery of future functionality?" then any refactoring is, to use Drucker's term, merely "polishing stones."*

• • • • • • • • • • • • • • • • •

*Re-reading these words I wrote in 2004, quoting Drucker from the 1950s, it is easy to see how prophetic his words were. In this second decade of the twenty-first century, we have come to observe that Agile is often pursued for the sake of itself; that "clean code" is pursued for its own sake. There are few, like Joshua Kerievsky, in the Agile community who are willing to put an economic framework around technical decision making and the acceptable effort to invest in code craftsmanship.

Drucker on Adaptive versus
Plan-Driven Processes
Tuesday September 7, 2004

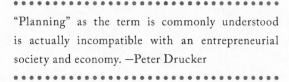

"Planning" as the term is commonly understood
is actually incompatible with an entrepreneurial
society and economy. —Peter Drucker

PETER DRUCKER DOESN'T LIKE PLANNING MUCH. HERE IS WHAT HE HAD TO SAY in 1967, writing in *The Effective Executive*:

> Most discussions of the knowledge worker's task start with the advice to plan one's work. This sounds eminently plausible. The only thing wrong with it is that it rarely works. The plans always remain on paper, always remain good intentions. They seldom turn into achievement.

Drucker also had something to say about adaptive process versus plan-driven processes back when he wrote *Innovation and Entrepreneurship* (Harper & Row, 1985). He completely predicted the shift to Agile software development methods and adaptive, just-in-time planning.

"'Planning' as the term is commonly understood is actually incompatible with an entrepreneurial society and economy . . . innovation, almost by definition, has to be decentralized, ad hoc, autonomous, specific and microeconomic."*

Drucker's argument is based in the idea that in a knowledge worker economy, a competitive edge—a differentiator—is achieved by bringing new ideas to market faster. This is similar to combining Marvin L. Patterson's concept, that product design is a process of information discovery, with Donald Reinertsen's observation that design-in-process is perishable. Drucker was basically arguing this in 1985. He had worked out that innovation is a knowledge-creation product of knowledge workers and that such knowledge has a half-life. To him, this meant that the nature of the economy was changing and that the paradigms we had used to manage the old economy of mass production were obsolete.

* * * * * * * * * * * * * * * * *

* Once again Drucker shows that he inherently understood that knowledge work is a complex activity that requires an adaptive and experimental approach to manage it.

Drucker on Teamwork

October 8, 2004

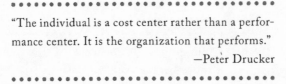

"The individual is a cost center rather than a performance center. It is the organization that performs."

—Peter Drucker

IN THE AGILE MOVEMENT, WE TALK A LOT ABOUT THE POWER OF TEAMWORK, and Agile methods spend a lot of energy on team working and on organizational structures and communication plans for team working. Peter Drucker agrees. Writing in *Management in a Time of Great Change* (Harper & Row, 1995) he says:

> In the knowledge society, it is not the individual who performs. The individual is a cost center rather than a performance center. It is the organization that performs.

I've mentioned before why individual measurement is bad (see chapter 2, Management Misdirection and Why Good Managers Still Matter to Agile Development), but Drucker seems to go further. He thinks it is pointless.

Why Individual Measurement is Bad

April 14, 2004

• •

Individual measurement is de-motivational. Poorly motivated developers underperform. Good leaders motivate. Bad ones don't. Managers who measure individuals aren't good leaders.

• •

I HAD DINNER WITH JOHANNA ROTHMAN, HERE IN SEATTLE, ON MONDAY evening. We got to talking about the problem of measuring developers individually. She sees this often with her clients. It typically comes just after an organization is beginning to stabilize. It is possible to measure velocity, to use the result to estimate a team's capacity, and then to limit the input to a rate that can be processed by the current team. Now that it is possible to measure, the boss starts thinking about how it might be possible to increase productivity. I had seen this when I was a manager at Sprint PCS. The newly appointed Senior Director for Engineering, once he realized that we had such wonderful data, metrics, and indicators, immediately wanted to use that information to assess individual performance. I refused. Our relationship deteriorated. It didn't end well.

Most managers believe that software development productivity is closely related to individual ability. Studies from the 1960s and early 1970s gathered data on so-called "programmer productivity." From this we learned that productivity differences can be huge. Sackman et al.[1] reported this in the late '60s. So managers start to think about how to identify the weaker links so that they can (they hope) eliminate them.

THIS IS SO WRONG!

You simply cannot measure developers individually. Why not?

Software development is a knowledge-work business. Knowledge is about information. The more knowledge and information available about the problem domain, the better off we all are. When you start to measure developers individually, you incentivize those developers to hoard information. Why share when you can be rewarded for keeping it to yourself? Why share when doing

---

1. Sackman, H., W.J. Erikson, and E. E. Grant. 1968. "Exploratory Experimental Studies Comparing Online and Offline Programming Performance." *Communications of the ACM* 11, no. 1 (January): 3-11.

so allows your colleagues to go faster? Whether it is information about the use of a language, a skill in UML, knowledge of unit-testing techniques, domain subject-matter knowledge, or just plain use of the IDE, it is all useful knowledge that can help a team go faster. Knowledge sharing is a key to success. To elevate the practice of software engineering, you need to encourage more—not less—knowledge sharing.

Individual measurement is anti–team working. Groups of individuals typically perform poorly compared to a good integrated team.

Individual measurement is also de-motivational. We live in a world where geeks grow up to be conspiracy theorists. For their protection, their government spies on them in many ways; so do their employers. Commercial entities spy on them, too, particularly via their web site usage. Their privacy is invaded daily. The last thing they need at work is a boss who spies on them and then punishes them for under-achievement—however that is defined. Knowledge-worker productivity is directly related to motivation. If the developer is a little coding machine, that machine works harder when pumped with the right motivation.

Individual measurement is de-motivational. Poorly motivated developers under-perform. Good leaders motivate. Bad ones don't. Managers who measure individuals aren't good leaders.

Developer As Executive

October 12, 2004

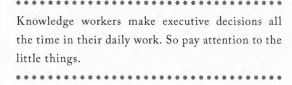

Knowledge workers make executive decisions all
the time in their daily work. So pay attention to the
little things.

WRITING IN *THE EFFECTIVE EXECUTIVE* (1966) PETER DRUCKER EXPRESSES
that all knowledge workers are de facto executives.

> I have called "executives" those knowledge workers, managers, or indi-
> vidual professionals who are expected by virtue of their position or their
> knowledge to make decisions in the normal course of their work that
> have impact on the performance and results of the whole.

By this token, according to Drucker, all software developers are executives. Why is this true? And what does this mean for software-development
management?

Every line of code a developer writes affects the performance (functional
or non-functional) of the product or system in development. For a software
product company, the competitiveness of the business relies on its product's
performance. More than that, the competitiveness of the business requires that
the software-development group be capable of responding to market demands
with appropriate functionality, quality, and timeliness. Every design decision potentially affects the ability of the business to respond to those market demands.
If developers are simply left to design, develop, and test on their own without
supervision, individual developers are making "executive decisions," on their
own, every day, in an open loop.

It isn't good enough to demonstrate that code meets the functional requirements, because working code disguises the many executive decisions that were
made while producing it. The effects of these executive decisions might not be
felt for many months to come. Cruft in the code can lead to long lead times,
poor quality, and dissatisfied customers. The introduction of design and code
inspections based on established guidelines and best practices effectively elevates
the executive decisions to a group or communal level. The guidelines should
encode the executive decisions that have been made concerning tradeoffs; for

example, flexibility versus simplicity in design. The guideline documents reveal the executive decisions made by the technical team so that management can agree that the decisions are properly aligned with the strategic direction of the business. The inspection process is there to verify that executive direction was carried out correctly.

Agile methods, such as Feature-Driven Development, can remove executive decision making from the individual: Such methods ask for designs to be completed by feature teams and for code to be reviewed by a feature team; a Chief Programmer is held accountable for the execution of the process and the quality of the product. Individual developers are not empowered to make executive decisions on their own—rather they are enabled to contribute to them as part of a team. In this respect, management can rest a little easier knowing that some control is being exercised over the strategic direction and its alignment with tactical decisions.

The term "technical debt" has crept in to the vernacular of the Agile movement recently. Technical debt is the notion that code is in a state unfit for long term maintenance—it works now, but only just. In order to continue using and reusing the code in future iterations, some refactoring is needed. Technical debt is the stuff of executive decisions. When a decision is made to leave a cruft debt in the code base, it should be exposed and visible to management. It should be part of the guidelines; that is, if YAGNI[1] is your mantra, your guidelines should explain what that means and provide examples. Management can then decide whether the choices are aligned with the direction of the business and needs of the market. If cruft exists because developers were left on their own to create it, making executive decisions in an open loop because they didn't have the full, "big picture" of the business, its market and customers, or its strategic goals, ultimately the business will get what it deserves—beaten in a competitive market!

The message from Drucker is clear—knowledge workers make executive decisions all the time. So pay attention to the little things. Understand that each knowledge decision is an executive decision and put processes in place to control the quality of those decisions and ensure that they are aligned with the strategic direction of the business.

---

1. "You Aren't Going to Need It"

Personal Hedgehog Concept

November 27, 2003

IN *GOOD TO GREAT*,[1] JIM COLLINS PROPOSES HIS THEORY OF a *HEDGEHOG CONCEPT*. It's the magic strategic position that enables firms to transform from good to great. I'd like to propose that the Hedgehog idea can be used for personal careers as well as for corporate strategy.

First, let me explain the Hedgehog Concept. Borrowing a graphic from the book, the concept is the intersection of three ideas—what you are passionate about, what you can be best in the world at, and what drives your economic engine.

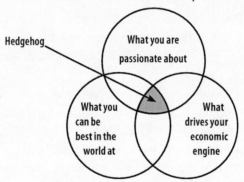

Figure 3.1   Jim Collins's Hedgehog Concept

To illustrate this, I'd like to use the example from the recent career of blogosphere cognoscento Cameron Barrett.[2]

I've met Cam on a few occasions and had the pleasure of dining with him twice while I lived in Kansas City. He had recently taken a job with the Wesley Clark presidential campaign, in Arkansas, a far cry from his web developer contracts in New York City. This led to criticism on his Camworld site, such as:

1. Collins, Jim. *Good to Great*. Harper Business, 2001.

2. This article was published with Cameron Barrett's consent

> Sorry Cam, but Clark won't win AND web-based campaigns are overrated. Yes, they build buzz, but it's miniscule and really has no impact on the real vote. Show me real data that disagrees with me. We can get all excited about this stuff, but it's mostly insignificant besides ego building.

Let's consider this criticism using Collins's Hedgehog Concept.

Those who know Cam or have followed his blog for many years, know that he is passionate about web standards, UI design, and web development; but they also know that he is deeply passionate about politics—particularly traditionally Democratic liberal issues. So the passion part of the Hedgehog diagram is easy—politics!

What drives Cam's economic engine? For years he has been paid to build state-of-the-art websites with a focus on open standards, good UI design, and high performance-to-cost ratios.

That leaves "What you can be best in the world at?" It seems to me that the market niche for building a winning online political presence is wide open.

Therefore, I feel that it's irrelevant whether Wesley Clark wins or loses. Cam can only be a winner. He will come out of the experience with a Hedgehog Concept that positions him as the world's leading developer of political campaign websites. Now, that is something at which he could make a career for the rest of his working life.[1] The web can only get more and more important as a political tool over the next 20 years.

---

1. As it happened, Wesley Clark withdrew from the primary campaign and Cameron went on to work on John Kerry's presidential campaign. Ultimately, the experience burned him out. He quit working for Kerry before the general election, and he did not go on to pursue a personal hedgehog in the world of online political campaigns.

Inspired by Deming

THE WORK OF W. EDWARDS DEMING HAS BEEN A MAJOR SOURCE OF INSPIRATION FOR ME AND IT HAS GREATLY INFLUENCED THE KANBAN COMMUNITY. IT IS DEMING WHO GAVE US THE TERM "CAPABILITY," WHICH WE PREFER OVER "PRODUCTIVITY" OR "PERFORMANCE." DEMING TAUGHT US how to understand that capability must be viewed in light of its variation and that it isn't a constant (or an average.) Deming gave us methods for studying and improving working processes. Most of all, Deming truly defined the concept of "respect of people" (within a system of work.)

The Agile and Lean movements both recognize a "respect for people" but I find that many practitioners fail to understand what this means. Deming taught us that to truly respect the people working in our business, managers must design a system that sets them up for success. To do so, there must first be understanding—an understanding of the capabilities of the system within which the workers are performing tasks, and an understanding of the demands set upon that system.

Deming's work also predicted results in software engineering. Studies done by Boehm (1981) that suggest the highest leverage for improving the effectiveness of software development is through improving management. Deming's 95/5 rule suggests that 95 percent of the observed capability of a system is determined by the design of the system, and that the differences in the capability of individual workers within that system affect only five percent of the capability. This result is counter-intuitive and generally elicits an emotional denial from talented software developers. With Kanban, I've been promoting the concept of understanding the system in operation, explicitly defining the policies that control that system, and rightly making it the role of managers to take ownership of those policies and the system design. Kanban training, therefore, is actually management training—its goal is to improve the quality of management decisions at all levels within an organization.

I've chosen to include a paper first developed for the Agile 2005 conference. Because I was in a sponsored speaking slot courtesy of my role with Microsoft at that time, I sought to present my work on MSF for CMMI® Process Improvement, demonstrating that Agile values and principles need not be sacrificed in order to deliver a process that is compliant with CMMI Maturity Level 3.

This work reflects my investigations of Deming's ideas at that time. I was still very much in the mode of collecting data and validating it against Deming's ideas. My work in 2004 and 2005 seems to have brought a focus on Deming to the Kanban community. This focus isn't always healthy. I've come to realize that software development work exhibits wide ranges of common-cause variation and that a narrow focus on reducing it may not be optimal from a customer-service perspective. It is also likely that control limits such as +/−3 sigma and the Shewhart method for calculating control limits[1] are not relevant in our world.

I have observed in the Kanban community a tendency for people to use lagging indicators, such as lead time, in control charts. Most Kanban tracking tools support this. It is arguable that such charts are of little use other than as historical report cards to reflect upon significant process changes, or to show an improvement in average lead time and predictability through reduced variation in lead time. The intent of control charts as a management intervention tool both to teach and to avoid Mistake #1 and Mistake #2 are lost.

So although this article is included for historical reasons and for its value in connecting the CMMI with Agile, the title of this chapter was carefully chosen. Choose to be inspired by Deming rather than dogmatically following his methods. Deming worked in a different era and, given his own nature, I am sure he would have sought to understand knowledge work just as he did manufacturing work. The results would likely have led to an evolution of his approach. The Kanban community has this opportunity today and I'd like to see where it takes us.

---

1. http://www.itl.nist.gov/div898/handbook/mpc/section2/mpc221.htm

What's Wrong with Conformance to Specification?

Friday, March 26, 2004

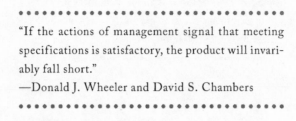

"If the actions of management signal that meeting specifications is satisfactory, the product will invariably fall short."

—Donald J. Wheeler and David S. Chambers

I FOUND THIS WONDERFUL QUOTE IN *UNDERSTANDING STATISTICAL PROCESS CONTROL* by Wheeler and Chambers. It's from the teachings of W. Edwards Deming. I feel it communicates the idea that we need a probabilistic approach to managing software development rather than the predominant twentieth-century deterministic approach.

The engineering concept of variation has the concept of meeting specifications.

. . .

Management has been trying the [engineering concept] since the beginning of the industrial revolution. After almost 200 years, the goal has not been met. The legacy of focusing solely upon conformance to specifications has been a lack of progress. There is no reason to believe it will be different in future.

. . .

Thus we come to the paradox. As long as management has conformance to specification as its goal, it will be unable to reach that goal. If the actions of management signal that meeting specifications is satisfactory, the product will invariably fall short.[1]

---

1. Wheeler, Donald J., and David S. Chambers, *Understanding Statistical Process Control*, 2 ed. Knoxville, Tenn.: SPC Press, 1992, pages 11–12.

Reduce Variation, Reduce Cost

September 16, 2003

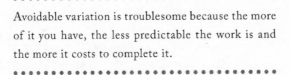

Avoidable variation is troublesome because the more
of it you have, the less predictable the work is and
the more it costs to complete it.

WHY IS VARIATION IMPORTANT? BECAUSE WITHOUT VARIATION IN OUR WORK, that work would be of no value. Knowledge work has a high level of uniqueness to it. So variation is inevitable. However, a lot of the observed variation in knowledge work is avoidable. This avoidable variation is troublesome because the more of it you have, the less predictable the work is and the more it costs to complete it.

To meet project promises and deliver software on time, schedules must be buffered against common-cause variation. This requires an understanding of a team's past performance and a prediction that their performance will be similar in the future. For example, if historically it has taken 2.2 days +/– 1.0 day to complete each feature, it is likely to take a similar amount of time in the future.

To be sure of delivery, a manager would need to buffer 1.0 day over the mean 2.2 days for such a feature. A whole project of similar features would require a buffer determined from the square root of the sum of the squares of those individual buffers. For 50 features, the required project buffer would be 7 days. The total project length would be 110 days + 7 days = 117 days.

Observed variation might be reduced if the manager can increase the team's skill level: Mentored training in analysis and design, capturing common elements as patterns, and encouraging collaborative working so that the capability of less experienced team members quickly normalizes with the team's best contributors could boost the team's skill level and cause observed variation to fall. Say it fell to only +/– 0.4 days. The result would be that the required buffer for 50 features would be only 3 days. This saves 4 days of effort. The smaller the quantity of work to be delivered, the greater the benefit realized from reduced variation.

Reducing variation, through better, more repeatable structured analysis of software deliverables improves predictability and saves money by reducing the total time it takes to deliver projects.

* * * * * * * * * * * * * * * * * *

* Writing in 2012, some nine years after this article was written, I find great irony in this piece. The Agile community is reporting an ever-increasing need to reduce the variability of user stories into small, fine-grained units with a narrow variation in size, complexity, and level of effort. The reason for this is the pressure exerted by ever-smaller time-boxes for iterations (or Sprints, in Scrum). This pressure comes from a need to improve agility and to be more responsive to customer needs.

However, after a decade of Agile and a focus on unstructured requirements' definitions using user stories, the ability to write controlled, fine-grained requirements with a narrow spread of variation in size and complexity has largely been lost. The irony is that more structure analysis methods from the 1990s will actually help with this situation, but such methods have fallen out of fashion, and almost a generation of software developers have graduated college without learning these skills.

On the positive side, Kanban actually helps this situation. By decoupling the cadence of queue replenishment and selection (prioritization and planning) from lead time and delivery cadence, the pressure of short time-boxes is removed. Kanban enables the improved agility required by the business without having to introduce advanced (and, ironically, old and unfashionable) structured analysis methods to provide small, fine-grained requirements to feed a team using an Agile method with tight time-boxes.

Lightning Strikes Twice

Monday, September 20, 2004

· ·

Something that hasn't happened in seven years suddenly happened twice in one day. That's clearly a
special-cause variation. So Shewhart would tell us
to go look for the assignable root cause.

· ·

I TALK A LOT THESE DAYS ABOUT STATISTICAL PROCESS CONTROL AND THE
concepts of common-cause and special-cause variation. Sometimes the statistics
scare people off. However, you really don't need to be an expert in statistics or at
using mathematics to recognize special-cause variation. Take today for example:

I got up with a plan to bike to work. I got ready, packed my rucksack with my
towel (no towel service in the Microsoft locker rooms any more) and my clothes
(no permanent lockers available in my building), and I actually left my garage
on my bike. Ptup, ptup, ptup, ptup, ptup, ptup. Arrrgghh! A flat! How did that
happen? Slow puncture. I found a staple stuck in the tire and assumed that was
the cause. I quickly reopened the house and the garage and changed the inner
tube. I set off again, now much later than I wanted to be. I eventually got in to
work an hour and ten minutes later, around 8:45 a.m. The last two miles were accompanied with a ptump, ptump, ptump sound. Strange, I thought, I don't recall
anything that would have flat spotted the rim. Perhaps it's a bulge in the tire?

Hurrying to shower and change, I didn't investigate further. Then I worked
late because I didn't get logged on until 9 a.m. I went down to the locker room
about 5:45 p.m., got changed, and walked out to unlock my bike. Aaarrrggghhh!
Another flat!

Lightning doesn't strike twice—as the saying goes. In seven years of own-
ing that bike, I have never had a puncture while riding around the city. It's a
mountain bike. It's tough and has big, thick tires. Riding in the Malaysian jungle,
yes, I've had flats, but never in the city, not in seven years! So what is the sigma
calculation on that? Who cares? What is blatantly obvious is that something
that hasn't happened in seven years suddenly happened twice in one day. That's
clearly a special-cause variation. So Shewhart would tell us to go look for the
assignable root cause.

Closer inspection of the tire revealed the problem. The tire wall had worn through on about a two-inch stretch. It wasn't protecting the inner tube, and clearly, the tube had snagged against the rim. Closer forensic inspection showed that the hole in the tire lined up with the frayed lining. I repaired the puncture and crossed my fingers. I returned home, relieved that there were no further incidents, but much later than I wanted, at 7:15 p.m. I inspected the other inner tube from the morning. Yep, there it was—a pin-sized hole exactly in the same place as the more recent one this afternoon. The root cause had been identified. Time for a new tire—and a couple of new tubes!

Don't Apologize, Be on Time!

September 1, 2003

• •

The accident and the fire are known in the quality-assurance field as special-cause events. . . . To buffer for special-cause events might allow for a face-saving delivery on time, but it is very expensive.

• •

"Don't apologize, be on time." That advice was given by Sean Connery to Wesley Snipes in the movie, Rising Sun. It was one of the few true insights into Japanese culture in an otherwise poor movie.

Hence, it was with some surprise that I stood on the sidewalk outside my office last Friday waiting for my (Japanese) wife, who was late. When I called her she was "two minutes away." Ten minutes later she showed up, with only 20 minutes to take me to an appointment that, at best, was 20 minutes away.

It turned out that she had set out in plenty of time, but she'd gotten stuck behind a truck that had grounded, turning off Ballard Bridge onto Nickerson, in Seattle. The unusually tight corner features a high concrete curb designed to protect pedestrians and cyclists from traffic accidents. The truck had beached itself on this curb. On the return trip we hit the same jam, as emergency services tried to clear the mess with a heavy-duty tow truck. In the end we were 10 minutes late. My wife's journey, which would normally have taken 20 minutes each way, actually took over an hour. Is it reasonable to attribute blame in this kind of situation?

There is considerable variation possible between my home and my office. The drawbridge can be up—typical delay, 5 minutes. There are numerous traffic lights. The level crossing can be closed for a train—typical delay, 10 minutes, but an alternative route is available. A typical journey takes 20 minutes; a good time is 15 minutes. If most traffic lights are red, the bridge is up, and the train is coming, it might take 30 minutes. However, if there is an accident (as happened in this case), the journey could take much longer.

An accident like this is known in the quality-assurance field as a special-cause event. The others—train, lights, bridge—are common-cause events. To buffer for special-cause events may allow for a face-saving, on-time delivery, but it is very expensive. Most estimates would be way over and projects would look far too expensive. For the "collect your husband and take him to an appointment" project, a reasonable estimate for the round trip might have been 50 minutes. This would have absorbed all reasonable common-cause variation with an almost 100 percent certainty of completing the job on time.

The lesson from this is that any reasonable senior manager should expect line managers to buffer common-cause variation appropriately, while accepting that unforeseeable special causes can still make a delivery late. In the event of unforeseeable special-cause variation, they should accept the outcome as a consequence of an appropriate risk-management choice and avoid attributing blame or fault.

Drive Out Fear!

Thursday June 17, 2004

• •

We all know that non-conformant quality is the norm in the software development world. So there is reason to be fearful.

• •

IN DEMING'S THEORY OF PROFOUND KNOWLEDGE[1] AND HIS 14 POINTS FOR Management,[2] he emphasizes the importance of driving out fear from an organization. Driving out fear is incredibly important for effectiveness. Deming underpinned his Theory of Profound Knowledge in the statistical methods of process control. He observed, "Some of the greatest contributions from control charts lie in areas that are only partially explored so far, such as applications to supervision, management, and systems of measurement. . . ." (Shewhart, 1986). In other words, Deming liked the idea that someone would come along at a later date and apply his theories to other areas, such knowledge work, and specializations like software development.* The trick for us will be to take inspiration from Deming, and leverage his ideas, while not blindly copying them from the manufacturing realm into knowledge-work fields.

Deming believed that understanding variation is the vital ingredient in driving out fear—the second element in his Theory of Profound Knowledge. Managers must understand variation and know how to separate common- (also known as chance- or systemic-) cause variation from special- (also known as assignable- or external-) cause variation.

Wheeler's Four States of Control,[3] and in particular the Threshold State, helps us to understand how it is possible to reduce fear in an organization. The Threshold State says that the system (the process of software development) delivers some non-conformant quality; that is, the project is late, or over-budget, or dropped scope, or has a higher than acceptable defect count, or perhaps all of those. The essence of "non-conformant quality" is that there was no assignable-cause variation but still we failed to deliver a satisfactory result for our cus-

1. http://www.maaw.info/DemingExhibit.htm

2. http://www.hci.com.au/hcisite2/articles/deming.htm

3. *Understanding Statistical Process Control* (Longman Group, 1990)

tomer. We all know that non-conformant quality is the norm in the software development world. So there is reason to be fearful. Unhappy customers are likely to lead to overreactions from our managers. How can we drive out fear in a world where non-conformant quality is the norm?

Figure 4.1 Donald Wheeler's Four States of Control

Managers must be responsible for educating staff on variation and helping them understand its effects. There will be fear at the staff level only if assignable-cause variation has been inadequately addressed through risk-management planning. Even then, the only fear should be of the impact of the event, and not of repercussions from managers. As described in "Don't Apologize, Be on Time," there is no point in assigning blame for assignable-cause variation that is beyond the control of the team, the managers, or the organization. And there is never a basis for assigning blame for excessive common-cause variation. Large amounts of common-cause variation are the norm in software development. If we can teach managers to see it for what it is, we can teach them to act appropriately. The result will be less fear in the workplace.

Managers, on the other hand, must carry the burden for common-cause variation. It is all too easy for managers to deflect accountability and falsely make claim to an assignable cause for variation that exceeded the bounds of the prediction in the project plan. How many staff live in fear that their manager will blame them for something over which they had no control? Most current software development methods suffer from wide spreads of common-cause variation. This comes from variation in the nature of requirements, variation in the solicitation method, variation in the analysis method used to break requirements into distinct traceable chunks—such as use cases, user stories, features, and so forth—and variation from the inherent technical complexity that each requirement might create.** This means that buffers in plans have to be large or the plan is at risk. Even if these projects are profoundly successful at eliminating special-cause variation through use of techniques such as the "container" concept described in the Scrum method, which is said to create a "punctuated equilibrium," at best they exist in the Threshold State due to underestimation of the common-cause variation.

• • • • • • • • • • • • • • • • •

*I believe Deming would have been fascinated and amused by the experimentation with control charts that is now happening in the Kanban community.

• • • • • • • • • • • • • • • • •

** Writing this in 2012, I have amassed considerable evidence to suggest that often 40 to 60 percent of the detail in requirements is missed on initial analysis. I referred to this phenomenon as "dark matter" in *Agile Management for Software Engineering*. Dark matter ratios of up to 200 percent have been observed. Variation in completion rate customer-valued requirements can often vary by a factor of two times above and below a mean. Variation in lead times to complete work can vary as much as five times above the mean time to complete. Almost all of this variation is inherent common-cause variation.

Managers can drive out fear by accepting responsibility for the system of software development and accountability for non-conformant quality. They can reduce their own personal risk by gathering data and reporting it transparently. By learning to recognize and report which quadrant of the Wheeler chart their system of software development is operating within, a manager can eliminate fear from the staff and increase the likelihood that they, as a team, can bring the process to the Ideal State over time. Only then can they start to use "Quality as a Competitive Weapon."

Quality as a Competitive Weapon

June 3, 2004

• •

A project in the Ideal state has eliminated assignable-cause variations and has reduced chance-cause variation sufficiently . . . to set appropriate expectations and, as a result, has delivered conformant quality—something that meets customer expectations.

• •

WHEELER'S FOUR STATES OF CONTROL MATRIX (SEE DRIVE OUT FEAR! IN chapter 4) tells you almost all you'll ever need to know about the Six Sigma approach to continuous improvement. The rows show the difference between assignable-cause* and chance-cause variation (these are better known by Deming's terms—special-cause and common-cause variation). However, Shewhart's original terms are perhaps more useful to us when considering management of knowledge work.*

• • • • • • • • • • • • • • • •

* I chose to use the terms assignable-cause and chance-cause in my book, *Kanban: Successful Evolutionary Change for Your Technology Business,* as I believe they better communicate a common understanding of the concepts.

Assignable cause. Something that has an assignable cause is identifiable; you can point to it. As such, an assignable-cause problem that has the possibility of occurring is a risk, and should appear on a risk-management plan. An assignable-cause problem that has occurred should be recorded in an issue log.

Chance cause. Variation that is endemic or systemic to a process and cannot generally be identified as having a root cause is chance-cause variation. Chance cause is the idea that Feature 167 on the backlog took 1 hour and 40 minutes to design collaboratively with the team, whereas Feature 168 took 2 hours and 10 minutes to design using the same people and a similar approach. The time it will take to design any one feature cannot be predicted with accuracy.

The columns on Wheeler's matrix are named for quality, but Wheeler's meaning is more general than most people would associate with the term "quality." Conformant quality means that customer expectations were delivered. From a project-management perspective, this might mean: all functionality, delivered on time, with a defect level below 20 bugs per 100 features.

Four States of Control

The Chaos state denotes that assignable-cause variations are present. The delivered quality is non-conformant, implying that we were unable to give the customer what they expected and that we suffered from both common-cause variation and assignable-cause variations. This might indicate that our estimation method produced an inaccurate estimate (chance cause) and that unforeseen things occurred, causing delay and further impact on actual delivery against the plan (assignable cause).

The **Brink of Chaos** state is when, despite assignable-cause variations, we still deliver something acceptable to our customer. Wheeler suggests that if this is the case, we got lucky!

In the **Threshold** state, we have eliminated assignable-cause variations, but our chance-cause variation is still not under control. This means that we fail to deliver something satisfactory to our customer, but that the failure is entirely due to the method we are using and our inability to perform it consistently enough or to set expectations accordingly.

Finally, a project in the **Ideal** state has eliminated assignable-cause variations and has reduced chance-cause variation sufficiently, or understands it adequately, to set appropriate expectations, and, as a result, has delivered conformant quality—something that meets customer expectations.

How do you use this framework for competitive advantage?

It all depends on what you define as "conformant quality." Have you ever considered what it means when a world-class golfer says that he is "rebuilding" his swing? At that level of golf, a player can hit balls all day with consistency in length and direction. The concept that assignable-cause problems affect play is rare: Clubs don't break, bees don't sting in mid-swing, low-flying aircraft are unusual. Most problems in golf are chance-cause problems involving the system of swing.

Referring back to Wheeler's model, the vertical axis that divides the Threshold State from the Ideal State can be defined competitively. For example, in professional golf it used to be good enough to drive 275 yards and land the ball in the fairway 75 percent of the time. Since the arrival of Tiger Woods, a top player now has to drive 290 to 300 yards and land the ball in the fairway around

80 percent of the time or better if they wish to be competitive. Tiger Woods has raised the standard, and from Wheeler's matrix perspective, he has shifted the *y*-axis to the right. He has made conformant quality harder to achieve. Players who can't meet the standard of driving 290 to 300 yards with at least 80 percent of them in the fairway are in the Threshold State—they don't deliver conformant quality, which means they don't win any tournaments.

In order to move their performance from left to right on the matrix, they need to change their swing—their system of hitting a golf ball. They need to change in order to reduce the chance-cause variation (improve their directional consistency) while delivering the new, higher standard of conformant quality (more power in the swing, more energy transferred into the ball, more distance from the tee). In golf, Tiger Woods moved that bar to the right, and in doing so he pushed his competitors into the Threshold state.

Application to Agile

So, consider what Agile methods are doing for software. Processes like Scrum try to eliminate the impact of assignable-cause variation through the communication mechanism of daily standup meetings. Impediments should be identified quickly and run to ground before they have an impact on performance. Ensuring this is one of the responsibilities of the scrummaster. However, in order to reduce chance-cause variation significantly, we need better technical practices and better, more consistent implementation of those practices. Code craftsmanship and test-first methods are just one approach. Better analysis mechanisms and better ways of sizing functionality and assessing complexity will also produce improvement. To get to the Ideal State, we need to have methods for codifying this. This is where Peter Coad's work with Domain Modeling in Color[1] and the Feature Template for defining requirements really helps. Mike Cohn's work to codify user story writing[2] gives us a chance of moving eXtreme Programming implementations from the Threshold State to the Ideal State. It is only by doing both—addressing the chance-cause variation through better analysis, planning, feedback, and reflection, and addressing the assignable-cause variations through better risk management—that we will be able to reduce the occurrence of blocking issues. Better day-to-day management to eliminate them swiftly when they do occur is how we will move a project into the Ideal State. Businesses operating in the Ideal State can use their conformant quality as a competitive weapon, just like Tiger Woods does!

1. *Java Modeling In Color With UML: Enterprise Components and Process* (Prentice Hall, 1999)

2. *User Stories Applied: For Agile Software Development* (Addison-Wesley, 2004)

Adapting Deming's Work to Explain
Agile to a Traditional Audience

Adapted from "Stretching Agile to fit CMMI Maturity
Level 3," Proceedings of Agile 2005, June 2005

• •

Deming's thinking is very well aligned with the
Agile Manifesto and the practices of many Agile
methods.

• •

Introduction

Back in August 2004, I started to look at developing a software development
system that would be compatible with the Software Engineering Institute's (SEI)
Capability Maturity Model Integration (CMMI) (SEI 2002) at Maturity Level 3.
At that time, I did not envisage that the final solution would be seen as an Agile
method. I assumed that the CMMI world and the Agile world were like oil and
water. CMMI and its predecessor, the Software CMM (SEI 1993), were largely
associated with the government-contracting aerospace and defense businesses—
a world of large, long-duration programs on often life-critical systems with government requirements for auditability and traceability.

Many software developers are suspicious of process generally. Process often
gets in their way and slows the pace of software development to a frustrating
level. In their minds, the CMMI is associated with such process initiatives. I arrived at the start of the MSF for CMMI® Process Improvement method project,
for Microsoft, with similar prejudices.

My interpretation of the essence of Agile software development was that it
is enabled by trust—trusting developers to do the right thing, and building trust
with customers through frequent delivery and attention to feedback. This provides the slack and empowerment needed to let good things happen. On the other
hand, the government aerospace and defense industries seemed to lack trust and
to rely on verification through audit—industries in which nothing happened
without committee review and approval. And approval was required frequently
throughout every project's lifecycle. Everything must be traced bi-directionally
in a fine-grained manner to facilitate auditing. In addition, the empowered,

self-organizing nature of Agile software development seemed opposed to the plan the plan, plan the work, work the plan, command-and-control approach that seemed to be called for in the CMMI. It seemed that a method compatible with the CMMI model needed to be a big planning up-front approach.

Some previous attempts to integrate Agile within a CMMI framework (for example, Alleman 2003), essentially offered point solutions, such as the ability to use some aspects of eXtreme Programming within an otherwise command-and-control, big planning up-front, non-adaptive environment. This seemed to underscore the difference in philosophy between the two worlds. What I sought to achieve was a full lifecycle method that was truly Agile but that met the requirements for the CMMI (at least to Maturity Level 3). The SEI (Paulk 2001, 2002) has sought to offer up such guidance, most notably this year (Konrad 2005). However, this guidance has tended to be focused on point solutions, such as XP, or high-level, philosophical guidance that is difficult to implement in a general, full lifecycle sense. The existing literature has surmised that CMMI and Agile were compatible but failed to provide specific details or examples.

I achieved my goal for an Agile CMMI method by stretching the MSF for Agile Software Development method to fit. The new method covers 20 of the 25 process areas (in CMMI version 1.2) with the following exceptions: Supplier Agreement Management, Integrated Supplier Management, Organizational Process Focus, Organizational Environment for Integration, and Organizational Training. Process areas in scope within Maturity Levels 2 and 3 on the CMMI model are covered in full, and those at Maturity Levels 4 and 5 have about 50 percent coverage.

What made this possible? What was the bridge between the CMMI world and the philosophy of Agile software development? The answer lies in the teachings of W. Edwards Deming. Deming's Theory of Profound Knowledge, based on an understanding of natural variation in processes and its analysis with statistical process control, was the key to unlock a full lifecycle Agile CMMI approach.

Deming and Agile

W. Edwards Deming's underlying philosophy of management is based on the concept of feedback—the Deming Cycle of Plan-Do-Check-Act (PDCA). Deming uses his feedback cycle to drive change based on objective statistical analysis. In addition to hard objective techniques like statistical control, Deming also introduced softer subjective guidance based on systems thinking with his 14 points for management (Deming, 1982). Of the 14 points, those of most interest to the Agile community are:

3. Cease dependence on quality control to achieve quality; instead focus on quality assurance throughout the lifecycle.

4. Build trust and loyalty with suppliers.

. . .

6. Institute training on the job.

7. Leadership.

8. Drive out fear.

9. Break down barriers between departments.

. . .

12. Remove barriers to pride of workmanship; focus management on quality rather than on production numbers.

Deming's thinking is very well aligned with the Agile Manifesto and the practices of many Agile methods. Number 3 could be compared to the practice of test-first development, for example. Agile is all about building trust between developers and customers (no. 4). Practices such as pair programming encourage on-the-job training (no. 6). Support for concepts such as sustainable pace, and transparency of reporting within self-organizing teams, drive out fear (no.8). Generalist roles and collaborative team-working break down barriers between departments (no.9). Short, iterative delivery cycles—where everyone on the team gets to see finished work and demonstrate it to the customer every few weeks— remove barriers to pride of workmanship and focus everyone on quality (no.12). According to Deming, that quality is defined by the customer, and not by the specification (that is, change must be embraced if that is what the customer wants). This is very Agile.

Deming taught the notion that quality is conformance to process rather than conformance to specification. Rather than measure conformance to specification using quality-control departments, it is better to build quality assurance into the process and measure conformance to process. The implication is that if you are doing things the right way, you will produce the right things, and the customer will be happy. It is second-order management—management of the process.

Theory of Profound Knowledge

The Theory of Profound Knowledge tries to provide an objective mechanism for measuring whether a process is under control—whether a team is doing the right thing. The theory is very simple. It seeks to understand the natural variation in tasks that workers are asked to perform and measure that variation. Using mathematics devised by Walter Shewhart (Shewhart, 1939), upper and lower control limits are set. If the process varies within the control limits, the process is said to be under control. Under control essentially means that the process is running as designed and it is not unduly influenced by external factors out of immediate management control. (This assumes that the process as designed is capable of running in a controlled fashion. This assumption is not without debate in the field of software engineering. Observed data from actual teams does suggest that software development processes are capable of running in control, though the observed spread of variation is often quite dramatic—on the order of two times above and below the mean.) When the process varies outside the control limits, it is said to be out of control. In other words, some external factor has affected the performance of the process and placed the quality of the output in jeopardy.

The Deming Epiphany

The CMMI is a model for process improvement that is intended to take an organization to a level where continuous improvement in productivity and quality is possible (Chrissis, 2003). This basic philosophy is based on the work of W. Edwards Deming (Deming 1982). However, the five-level model, which was introduced to the original Software CMM to enable the assessment of maturity in government contractors for United States Air Force contracts, was based on Philip Crosby's (1979) manufacturing model. Crosby's name is most associated with the definition of quality as conformance to specification (Crosby 1979). Hence, to measure quality, one must begin with a complete specification so that at the end the variance from the specification is measured and quality is assessed. For software projects, success is measured using this quality metric, interpreted as delivering a project on time, on budget, and with the agreed-upon functionality. The Crosby influence on the Software CMM is so prominent that it is not widely appreciated that the underlying philosophy and goal of the CMMI is to achieve continuous improvement—as described and taught by Deming.

The CMMI maturity model is mapped to Deming's thinking. Maturity Levels 2 through 4 are all about creating the organizational capability to eliminate special-cause variation, hence avoiding management mistakes Numbers 1 and 2 (see section after Figure 4.4), whereas Maturity Level 5 provides for continuous improvement through the gradual reduction of common-cause variation. It seemed to me that if the CMMI were truly rooted in Deming's philosophy, it must be possible to create a truly agile CMMI method.

It was this realization, in November 2004, that caused me to abandon the work done so far on MSF for CMMI® Process Improvement and to regroup by borrowing the emerging work on MSF for Agile Software Development and expanding upon it.

● ● ● ● ● ● ● ● ● ● ● ● ● ● ● ● ●

(I've come to appreciate more recently that a goal of eliminating special-cause variations is uneconomical and practically unreasonable. As we improve software development processes we must work on both common- and special-cause problems simultaneously. Hence, the observation that Kanban introduces Maturity Levels 4 and 5 practices immediately is perhaps no surprise. It seems that to truly achieve continuous improvement you must set this goal from the beginning. The CMMI model and how it has been used to drive managed change initiatives seems to imply that higher-maturity behaviors should not be introduced until later—once lower-maturity behaviors are well established and institutionalized. My experience with Kanban seems to suggest otherwise. Setting an expectation of a high-maturity outcome, right from the beginning, appears to produce better results much faster.)

CMMI Organization

The CMMI model can be viewed as either a staged representation or as a continuous representation. The model can be implemented in a staged fashion. This is the idea that an organization achieves Maturity Level 2 across all relevant practices, and then Maturity Level 3, and so on. It also can be implemented in a continuous fashion in which capability in an individual process area is measured. An organization can achieve a high capability level in a particular process area while holding only a capability Level 2 in other process areas. It is the staged method of assigning maturity levels rather than the continuous method based on capability levels that is associated with the American government contracting market and hence is best known in North America. The CMMI model actually consists of 25 process areas. Within each process area, a number of specific goals are enumerated. Each specific goal, in turn, has an enumerated set of specific practices. Each practice is set at an expected level that describes what should be done and what artifacts might be produced as a result. For example, Project Planning (PP) 2.1 asks us to "identify task dependencies" and expects a "Critical Path Method (CPM) or Program Evaluation and Review Technique (PERT)" (Chrissis 2003) chart as a result.

Many of these practices as documented would be unacceptable to an Agile software developer. For example, PP 1.1 sub-practice 2 directs, "Identify the work packages in sufficient detail as to specify estimates of project, tasks, responsibilities and schedule." It continues, "The amount of detail in the WBS at this more detailed level helps in developing realistic schedules, thereby minimizing the need for management reserve." It is all too easy to interpret this as a requirement for big planning up front. Project Monitoring and Control (PMC) 1.1 asks us to compare "actual completion of activities and milestones against the schedule documented in the project plan, identifying significant deviations from the schedule estimates in the project plan." This sounds like heavy planning—command and control—and is antithetical to the Agile concept of self-organization and postponed work allocation. Technical Solution (TS) 2.2 directs, "Establish a technical data package," and suggests artifacts such as "product architecture description, allocated requirements, product component descriptions, product characteristics, interface requirements, conditions of use," and so forth. Again, it is easy to interpret this as a requirement for heavyweight documentation. It also prescribes design before coding—which is abhorrent to the extreme test-driven development philosophy in which design is emergent through refactoring. In fairness to the CMMI, such specific guidance is defined as "informative" and is provided only to help readers understand what is expected.

In addition, the CMMI has a reputation for heavyweight documentation. Some literature (Ahern, 2005) advises that as many as 400 document types and 1,000 artifacts are required to facilitate an appraisal.

Achieving Agility with CMMI

A key to achieving more agility with the CMMI is to realize that the practices are primarily advisory or indicative only. To meet a CMMI appraisal, an organization must demonstrate that the goals of a process area are being achieved. This is done by identifying evidence of practices. However, the practices need not be the ones described in the CMMI specification. The organization is free to propose alternative practices and appropriate evidence. The appraisal team must then agree that this is an appropriate method to demonstrate the goal. This provides a process designer with a great deal of freedom.

As it turned out, MSF for CMMI® Process Improvement did not require any alternative practices except the practice and evidence for specific practice REQM 1.4, which calls for full end-to-end traceability of artifacts. This could be considered onerous for many Agile projects. The reason alternative practices weren't required has to do with interpretation of intent. When one practice asks for a plan to be made, and another asks for a commitment to a schedule, a budget, and a specification, people often assume that these commitments must be precise—that there is an underlying philosophy of conformance to plan and specification. When another practice asks for a review of planned versus actual and creation of a report detailing deviation from specification, there is an assumption that this is to audit quality as conformance to specification—and that any deviation is a failure in quality. Although this is a common understanding in many organizations practicing CMMI, it need not be true. It is acceptable to document the approach as conformance to process and the measurement mechanisms to be variation-aware and defined within common-cause variation limits. This is the key that unlocks the possibility of a truly agile CMMI implementation across the entire lifecycle.

Hence, plans and specifications need not define precise numbers and dates, but they may define a range of acceptable numbers and dates. So long as this is the communicated organizational understanding and it is acceptable to the customer, it is acceptable within the CMMI specification and appraisal scheme. For example, a plan based on historical velocity data of 1.7 scenarios per developer per week with a low-end variability of 0.8 and a high end of 3.4, might conservatively offer to develop 80 scenarios in ten weeks with ten developers, and define a stretch goal of 340 scenarios. A slightly more aggressive plan might use eight weeks with a two-week buffer for variation, and offer a minimum of 8 x 17 = 136 scenarios, with the same unlikely stretch goal of 340.

● ● ● ● ● ● ● ● ● ● ● ● ● ● ● ● ●

(I've come to realize that although an approach like this is logically correct, in practice it is hard or impossible to achieve. The spread of common-cause variation observed in the velocity data of real teams means that customers are seldom willing to buy into plans based on such wide spreads of variation. The Kanban approach—classifying work into different types and classes of service and offering different commitments (or service levels) for each type and class of service combination—provides a solution to this problem.)

Velocity and Queue Management

At the highest level of maturity, the CMMI calls for continuous improvement enacted under conditions of objective control. It is recommended that this be achieved using statistical control as promoted by Deming in his Theory of Profound Knowledge. It was desirable for us to deliver a method that met the CMMI Maturity Level 3, but also to lay the groundwork for a future version that could take adopters all the way to Maturity Level 5. The SEI firmly believes that the economic benefits of the CMMI are delivered at the higher maturity levels. At Microsoft, we believed that our customers would ultimately want to make the transition to Maturity Level 5, and we wanted to facilitate that with a smooth, incremental path using our existing process template. It was therefore necessary to consider Maturity Level 5 in our planning, even though we were focusing on delivering a Maturity Level 3 solution.

To follow a Deming approach to achieving CMMI Maturity Level 5, there must be a mechanism that allows monitoring of statistical control. Compatibility with Agile methods can be achieved by measuring velocity and using it for planning. Figure 4.2 shows the velocity of task completion (or the production rate of the process). This is very different from the more traditional project-management approach of estimating time-on-task for each task and then tracking planned versus actual performance. The latter merely measures variation in estimation ability, or lack of conformance to individual local commitment, whereas the former measures variation in the whole system of value delivery to the customer.

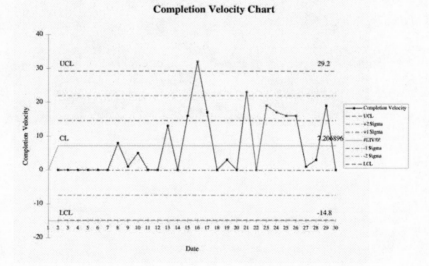

Figure 4.2   Completion velocity (or production rate) in a XMR control chart

The velocity measurement is not sufficient on its own to identify special-cause variation in the process. Shewhart called special causes "assignable causes." In other words, there ought to a specific assignable event that can be identified as the root cause. A record of project issues is a good source for identifying special-cause variation. The issue log stores the history of engineering requests for help. By implication, requests for help indicate potential or actual special-cause variation. There could be a special cause that is initially unidentified and therefore invisible. Management can choose to investigate or simply to shrug it off as unexplained. Clearly, more mature organizations are better at this root-cause investigation and exhibit processes under greater control.

We can track blocking work-in-process (WIP) by monitoring work queuing for processing. In *Agile Management for Software Engineering* (2003), I show how to use cumulative flow diagrams to do this (Figure 4.3), based on work by Reinertsen (1997). Monitoring the WIP level using statistical control (Figure 4.4) enables several things: WIP levels predict lead time, and therefore affect the future project schedule. WIP levels also can indicate blocking work items and identify where in the lifecycle the blockages are.

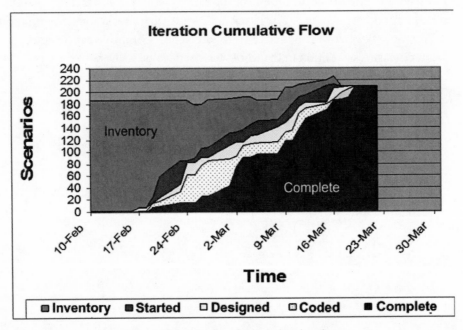

Figure 4.3 Cumulative flow of scenarios in an iteration

Deming suggested that there were only two mistakes a manager could make. He called these Mistake #1 and Mistake #2. Mistake #1 is interfering when everything is normal, that is, when the process exhibits statistical control. He called this tampering. When a process is under control, it should be left alone. Micromanagers tend to tamper because they react to fluctuations without considering the bigger picture. Hence, micromanagement often leads to Mistake #1. When a process is under control, the workers should be left to self-organize without management intervention. This concept is very compatible with Agile values. Mistake #2 is a failure to intervene when a process is out of control. A process gets out of control when something external affects it. In common language, we call these issues, or impediments. When an issue is blocking the value flow in an Agile software development process, everyone understands that someone (usually a project manager) must be assigned to eliminate it. Methods like Scrum (Schwaber 2001) make this explicit. Avoiding Mistake #2 is all about aggressive issue management and good risk management. A risk, after all, is simply an issue that hasn't happened yet.

WIP Inventory Chart

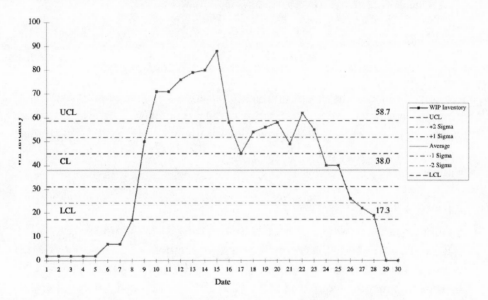

Figure 4.4. WIP scenarios shown in a control chart

The issue log should reflect the trends in WIP. It should contain details of assignable causes for observed special-cause variations in the data.

The issue log and blocked WIP can be used to assess the organization's capability to eliminate special-cause variation. Figure 4.5 shows the cumulative flow of issues in the issue log. Project managers work to resolve issues and close them out. On occasion, issues do not close fast enough, which results in blocked work-in-progress in the main project (or iteration) backlog. The overlaid line graph shows this. A lack of blocking work indicates a lack of special-cause variation, whereas the WIP Inventory and Completion Velocity charts show the extent of the common-cause variation.

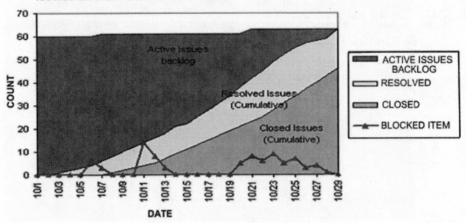

Figure 4.5. Issue log cumulative flow with blocking scenario inventory

An understanding of common-cause variation can be used for project planning and estimation. An understanding of the risk associated with special-cause variation can be used to pad the project plan with buffers (effectively, insurance) against schedule slippage or delivery of reduced functionality due to velocity slippage.

Iterative and Adaptive Planning

By defining a set of Agile metrics, but constructing them within a framework of statistical control, I laid the ground for an Agile method compatible with CMMI Maturity Level 5. What was needed to achieve the goal of an acceptably Agile method that met the requirements for CMMI Maturity Level 3 was an iterative and adaptive planning mechanism.

The solution was to provide a loose project plan that approximates the scope of a whole project and lays out a plan for a series of iterations that define approximately what will be developed in each one. The secret is to avoid big planning up front. I achieved this by defining a small series of critical to quality (CTQ), end-to-end scenarios, and storyboarding each one. As stated previously, the customer defines quality. It's important to gain an early, if somewhat approximate, mutual understanding of this. The customer must agree that these CTQ scenarios represent the broad intent of the product or service envisaged as the deliverable from a project. The CTQs are then assigned against iterations in the project plan. Hence, the project plan is very loose and approximate, but it is detailed enough to communicate to the customer the essence of what they will get at each stage in the project. It's important that the customer communicate what represents a suitable iteration cadence from their perspective. The natural business cycle is likely to dictate how often the customer can absorb new functionality and realize value from it. Hence, the customer determines the iteration length. A commitment to a project plan is therefore possible from all stakeholders. This commitment is a critical part of the CMMI expectations.

Detailed planning is then done at the start of an iteration, when a set of specific, detailed scenarios are fleshed out around the previously identified CTQ scenarios. Acceptance tests are written for each scenario, and together, the requirements and tests form the scope of the iteration plan. Scenarios are ranked using MoSCoW[1] rules (DSDM 2005). The metric for velocity and variation is used to define a minimum delivery level for the iteration, based on the low end of the capability range, and a stretch goal, based on the high end. "Must have" requirements are assigned against the minimum level, and "should have" or "could have" requirements against the stretch goal. All stakeholders then agree on the plan, and commitments are made.

During the iteration, the engineers are asked to focus continually on conformance to process, using the velocity and cumulative flow reports along with the issue log and the blocked WIP report. Attention will be paid to variation in velocity. A precise estimate of the scope that will actually be delivered can be made on a daily basis. It should continue to fall within the bounds of the minimum level and the stretch goal, or there should be an assignable cause to explain why it does not. Management attention can then focus on any problems and recovery actions initiated. By doing so, CMMI requirements for root-cause analysis and corrective action are covered.

Once again, at the iteration level, planning should remain loose. The plan for an iteration contains only a backlog sorted into "must have" requirements—which make up the minimum commitment level—and others, which represent a buffer for the natural variation in the process being used. The team is empowered to self-organize within the time-box and scope of an iteration plan. The plan is calculated using averages from historical data. When no history is available, the plan should be based on a best guess and monitored frequently using the reports shown in Figures 4.2 through 4.5. The plan can be adjusted accordingly. This only ever affects the first iteration of a new project with a new team.

● ● ● ● ● ● ● ● ● ● ● ● ● ● ● ● ●

It is worth noting and as I've commented similarly on earlier articles, that I no longer believe in this approach. Attempts using real data showed that the required

1. Must, Should, Could, Won't

buffers were too large to be acceptable to customers, managers, and other stakeholders. While logically correct, the approach is psychologically unacceptable. This lesson was a major driver toward the alternative approach used in the Kanban Method.

Stretching MSF for Agile Software Development

What the smaller Agile process lacked was some specific guidance and formal connection to some process areas within CMMI. In many Agile development methods, much of the infrastructure needed to develop software reliably is either implied or left to be resolved through a general-purpose issue management method. For example, if a lack of a configuration management environment is a blocking issue, the resolution would be to put one in place. The CMMI, based on decades of software engineering experience, already includes many of these aspects explicitly. Creating appropriate environments, commissioning appropriate tools, initiating projects with sufficient vision and communicating that vision with launch events, and planning for the realization of value through deployment—all are part of the CMMI model.

There is nothing in these practices that is particularly at odds with an Agile approach. The main difference is that they are spelled out explicitly.

It was therefore necessary to enhance MSF for Agile Software Development with additional activities to cover some aspects of the CMMI model. This was non-trivial. The result is a process template that is larger than a typical Agile process, with slightly more formality with respect to approvals, sign offs and audits. However, it has an Agile, adaptive core and is lightweight in comparison to traditional approaches to meeting the CMMI Maturity Model requirements.

Conclusions

CMMI process implementations are often associated with conformance to plan and low-trust environments with command-and control-structures. These require a big design up-front approach with auditing of conformance and, by implication, punishment for non-conformance. This generates fear and fear encourages greater focus on more detailed planning coupled with a blind willingness to follow such a plan. As a result, the customer's needs become disconnected from the work and customer-driven change becomes unacceptable because change must be reflected in the plan so that the result conforms to the plan. The effect is a great deal of non–value adding bureaucracy and rework.

This article shows that it is wrong to associate these undesirable software engineering behaviors with the CMMI. It doesn't have to be that way.

By embracing the teachings of W. Edwards Deming, and understanding their relationship to Agile principles and practices, it is possible to develop a truly Agile, full-lifecycle process that meets the requirements for all five Maturity Levels in the CMMI model. Specifically by using Agile metrics such as velocity, cumulative flow, and trends in open issues, we have designed planning and monitoring methods that embrace variation and allow for postponed, late-commitment, and adaptive iterative planning.

By doing so, the process embraces the Agile Manifesto's (Cunningham 2001) ideas, such as valuing *customer collaboration over contract negotiation* and *responding to change over following a plan.* In addition, by embracing variation and accepting it as inherent to the work, the people aspect of *individuals and interactions over processes and tools,* is embraced. Finally, via delivery in iterative cycles at an agreed-upon cadence acceptable to the customer, a focus on *working software over comprehensive documentation* is embraced.

References

(Ahern 2005) Ahren, Dennis M., Jim Armstrong, Aaron Clause, Jack R. Ferguson, Will Hayes, Kenneth E. Nidiffer, *CMMI SCAMPI Distilled—Appraisals for Process Improvement*, SEI Series, Addison Wesley, Boston, MA 2005

(Alleman 2003) Alleman, Glen B. and Michael Henderson, "Making Agile Development Work in a Government Contracting Environment—Measuring Velocity with Earned Value," Agile Development Conference, Salt Lake City, Utah, June 2003

(Anderson 2003) Anderson, David J., *Agile Management for Software Engineering—Applying the Theory of Constraints for Business Results*, Prentice Hall PTR, Saddle River, New Jersey, 2003

(Carroll 2000) Carroll, John M., Making Use—Scenario-Based Design Of Human-Computer Interactions, MIT Press, Cambridge, MA 200

(Chrissis 2003) Chrissis, Mary Beth, Mike Konrad and Sandy Shrum, *CMMI—Guidelines for Process Integration and Product Improvement*, SEI Series in Software Engineering, Addison Wesley, Boston, MA 2003

(Crosby 1979) Crosby, Philip B., *Quality Is Free*, Signet (Re-issue edition) 1980

(Crosby 1995) Crosby, Philip B., *Quality is Still Free—Making Quality Certain in Uncertain Times*, McGraw Hill Harvard Business School Publications, 1995

(Cunningham 2001) Cunningham, Ward et al, "The Manifesto for Agile Development," http://www.agilemanifesto.org/

(Deming 1982) Deming, W. Edwards, *Out of Crisis*, The MIT Press, Cambridge, MA 2000

(DSDM 2005) DSDM Consortium, MoSCoW Rules, Dynamic Systems Development Method, Version 3.0, http://na.dsdm.org/en/about/moscow.asp

(Konrad 2005) Konrad, Mike, and James W. Over, "Agile CMMI: No Oxymoron," *Software Development Magazine*, March 2005

(Paulk 2001) Paulk, Mark C., "Extreme Programming from a CMM Perspective," *IEE Software* November 2001, ftp://ftp.sei.cmu.edu/pub/documents/articles/pdf/xp-from-a-cmm-perspective.pdf

(Paulk 2002) Paulk, Mark C., "Agile Methodologies and Process Discipline," *Crosstalk*, October 2002, http://www.stsc.hill.af.mil/crosstalk/2002/10/paulk.html

(Reinertsen 1997) Reinertsen, Donald G., *Managing the Design Factory—A Product Developer's Toolkit,* The Free Press, New York, NY, 1997

(Schwaber 2001) Schwaber, Ken and Mike Beedle, *Agile Software Development with Scrum,* Prentice Hall PTR, Saddle River, NJ 2001

(SEI 1993) Paulk, Mark C.; Curtis, Bill; Chrissis, Mary Beth Chrissis, and Weber, Charles, Capability Maturity Model for Software Version 1.1, Software Engineering Institute, CMU/SEI-93-TR-24, DTIC Number ADA263403, February 1993.

(SEI 2002) Software Engineering Institute, CMMI® for System Engineering, Systems Engineering/Software Engineering/Integrated Product and Process Development/Supplier Sourcing, Version 1.1, Staged Representation (CMMI-SE/SW/IPPD/SS, V1.1, Staged), Carnegie Mellon University, Pittsburgh, PA 2002

(Shewhart 1939) Shewhart, Walter A., *Statistical Method from the Viewpoint of Quality Control,* Dover Publications, Mineola, NY 1986

Personal Hedgehog
Revisited

November 2, 2008

MORE THAN 4 YEARS AGO, I RIFFED OFF JIM COLLINS'S IDEA of a corporate Hedgehog Concept, with the previous post on the Personal Hedgehog Concept. It's proven to be one of the most popular blogs on AgileManagement.Net.

The original post used the career of Cameron Barrett as the example. At the time, Cam was pursuing his passion for politics, working to support the campaigns of Democratic presidential candidates—first, Wesley Clarke, and later, John Kerry. Recently I was asked to explain what the Personal Hedgehog Concept means to me.

Actually, I've been working on my own Hedgehog Concept for most of the past eight years.

First, what am I passionate about? For a long time, I've been passionate about the underperformance of the software-engineering profession and the low rate of success on software-development projects. In fact, I was so disgusted with the profession that I intended to quit almost ten years ago. It was thanks to Jeff De Luca, and the original feature-driven development[1] (FDD) project in Singapore, that I regained my enthusiasm.

So, what can I be one of the best in the world at? It's taken a while, but I started down the path to publishing and what we now call blogging in 1999, at the behest of Peter Coad and Jeff De Luca. Four years later, Peter was instrumental in assisting me with the publication of my first book, *Agile Management for Software Engineering* (Prentice Hall, 2003). I've continued to work at improving my ideas

---

1. http://en.wikipedia.org/wiki/Feature_Driven_Development

97

on managing both the software-engineering process and the knowledge workers who implement it. To do this, I continued to work as a practitioner in regular jobs managing software engineers at firms like Sprint and Motorola. That was until recently, when I formed David J. Anderson & Associates, Incorporated—a management-consulting firm based in Seattle, Washington.

So, what changed? Well, finally, I was able to realize my Hedgehog Concept. Finally, my skills with the software-engineering process and with managing and leading knowledge workers were in sufficient demand that they could drive my economic engine.

Let's be under no illusion! There is little to no premium in the market for good management in the software and IT industries. While great individual contributors often become independent contractors and earn high hourly rates, the same does not generally apply to managers. And although employers might be willing to pay a ten- to twenty percent premium for a decent person, often a great manager finds him- or herself earning far less than the top technical people on the team. This is despite the hard economic evidence[1] that management talent is generally what constrains the performance of software-engineering organizations.

So, I've known for a long time that I had to break out of working as a manager for other people and start my own firm. The question was, when? When would the timing be right? Finally in 2008, with a track record that included successful projects and teams at Sprint, Motorola, and Corbis, and with a catalog of intellectual property that included my contributions to FDD, the MSF for CMMI Process Improvement, and most recently, my contributions to Lean in software development and the innovations with

---

1. Boehm, Barry, Software Engineering Economics (Prentice Hall, 1981)

the Kanban method, I finally have sufficient recognition and respect in the industry for it to drive my economic engine.

Along the way, I've also resolved my own inner conflicts about whether I had taken the correct career path. I've finally come to realize that management and leadership are my real strengths, and that other things I enjoy are merely hobbies, like painting, art and design, and synthesizing those talents in user-interface and interaction design. It was, in fact, user-interface design that got me started down this road with my uidesign.net website. Recognizing in myself what I could be world class at versus the things that I can be merely good at has been the foundation of a new happiness in my life.

So, here we are! I'm having the best fun at work since I quit the games industry in the late 1980s, and I'm happier than perhaps I've been since leaving Singapore in 1999. Finding my Personal Hedgehog Concept has been at the root of this happiness. It's been a long slog—more than eight years— and a journey of personal discovery. Ultimately, it's been worth it. And now I am excited about the future. I intend to continue innovating in leadership and management of knowledge workers and helping teams deliver superior economic performance.

Goldratt and His Theory of Constraints

I T WAS ELI GOLDRATT'S WORK THAT INSPIRED ME TO QUESTION, "WHY DO AGILE METHODS WORK BETTER?" WHEN GIVEN THE CHALLENGE, IN 2000, OF SIGNIFICANTLY IMPROVING OUTPUT FROM A SOFTWARE DEVELOPMENT BUSINESS UNIT AT THE UNITED STATES TELEPHONE COMPANY, SPRINT, without significantly increasing staffing level, my boss handed a copy of *The Goal* to all his direct reports. He hoped that it would inspire us to find the bottlenecks in our organization and to improve them in an economically efficient manner. It was the start of a journey for me.

In 2000, we still called new, better ways of developing software, "lightweight methods." The term "Agile" would not be coined until the following year. There was little talk of an underlying theory of how these new lightweight methods worked. They just did! Many in the emerging community held superstitious beliefs about the appropriateness of these new processes. There were only a few people around, like Alistair Cockburn or Jim Highsmith, who were working on any underlying theories or explanations. There was talk of "ecosystems" and "fragility." Superstitiously, we were told to "do all 12 practices" or the method wasn't guaranteed to work.

For me, the Theory of Constraints gave us a model to look for an explanation of why Feature-Driven Development worked better than traditional methods. If felt that if I could show that it better exploited bottlenecks, we'd be onto something.

Why was this important?

Simply put, if Agile methods were to be widely adopted in large corporations, like Sprint, there would need to be compelling evidence to justify their adoption. Mainstream market businesses were not likely to adopt Agile on the basis of superstition or tribal behavior—evidence would be required, and that evidence would need an underlying framework or model on which to pin it.

What I didn't expect to find in Eli Goldratt's writings were the foundations of what we now call the Kanban Method—the foundations of an incremental, evolutionary approach to change and an explanation for why people resist change in their organizations and in their working practices.

Goldratt's Five Focusing Steps, and his observation that people resist change for emotional reasons, were the clues and inspiration I needed to look for an incremental, evolutionary approach to improving agility in software development organizations. At first, I was looking for bottlenecks. Later, I realized that there is so much variability in the nature of our work that bottlenecks can be hard to pin down. Controlling the variability in flow with a kanban system was a more fruitful approach. But there were many lessons to be learned from Eli before that. . . .

Lessons Learned from Eli,
#1: Small Batches

Thursday, June 24, 2004

• •

When you are struggling to keep a vehicle under
control at high speed, take your foot off the gas, and
slow down. We all know that things are easier to
control when they are moving more slowly.

• •

ON JUNE 24, 2004, ELI GOLDRATT GAVE ONE OF HIS VIABLE VISION TOUR
seminars in Seattle. I picked up a number of little insights from some of his
more subtle comments. This is the first of them.

Small Batch Sizes

Donald Reinertsen taught me that instituting small batch sizes (coupled with a
focus on quality) is often enough to bring in big results. In other words, forget
gathering data, identifying the constraint, and so forth; simply reduce the batch
sizes and focus on building quality in, rather than catching defects later, and
things will improve immensely—ergo, your client (if you are a consultant) will
be happy.

Eli Goldratt put it this way, "often reducing batch size is all it takes to bring
a system back into control." This ought to have been obvious to me—a trained
control systems engineer—because I learned it in college. To bring something
back inside the control envelope, simply reduce the amplitude of the signal. In
this case, the amplitude is the batch size in the process. In layman's terms, when
you are struggling to keep a vehicle under control at high speed, take your foot
off the gas, and slow down. We all know that things are easier to control when
they are moving more slowly.

In most human endeavors we reward control under high-amplitude conditions. We reward the fastest drivers, the fastest runners, the fastest mountain
bikers, skiers, speed skaters, and the list goes on and on. We intuitively know that
control under high speed is hard. Control under any high-amplitude signal is
hard, so we base measurements for graceful, subjective sports such as ice skating,

diving, most X-Sports—such as free skiing or BMX biking—for the style under the height of the jump, and the speed of the movement, or number of rotations. In industry and project management, it is the size of the project or iteration that represents the amplitude in the control signal. Hence, maintaining control with large batch sizes is hard. When something is out of control, it is, by definition, failing to deliver conformant quality—and the customer isn't happy.

Lessons Learned from Eli, #2:
Resistance to Change

Tuesday, June 29, 2004

• •

An organization running in a state of control is one that no longer needs heroes!

• •

AT HIS VIABLE VISION SEMINAR, IN SEATTLE, ELI GOLDRATT DESCRIBED SOME of the reasons why people resist change, and what it is about the culture of an organization that creates an environment that molds such people. I realized that he was talking about what Jerry Weinberg[1] has described as Level 1 and Level 2 organizations—the hero developer level and the hero manager level. The hero is cast in the role of firefighter, and he is the hero because he delivers. He delivers by putting out fires. As a result, he is rewarded for putting out fires and he is praised and admired by his colleagues as a champion firefighter. The more fires, the better practiced he becomes at putting them out and the more admired he becomes for putting them out. As a result, he measures his self-esteem by his prowess at putting out fires.

Hence, the hero firefighter learns to thrive on chaos. Chaos is the norm in the organization and the hero is the master of chaos—the one who parts the seas and delivers the team from the perils of chaos and non-conformant quality.

A hero does not want to move to a state of control because his or her self-esteem will drop when she or he is no longer praised for being a hero. An organization running in a state of control no longer needs heroes!

So there is a conflict. The organization wants to be under control and to deliver predictably, but some staff members thrive on chaos and their status in the organization depends on it.

How might we resolve this conflict? The senior management must start to reward people for behavior that is congruent with controlled performance, and they must build self-esteem around that behavior. The heroes must be coached

---

1. Weinberg, Gerald. *The Psychology of Computer Programming.* New York: Dorset House, 1998

and assisted to adapt to a new pattern of behavior—one that anticipates and absorbs uncertainty rather than one that heroically reacts to it.

● ● ● ● ● ● ● ● ● ● ● ● ● ● ● ● ●

As a further reflection on this story, I encountered such a situation in 2007—the hero group project manager who thrived on chaos and firefighting. Like Goldratt, I'd always believed that changes should be made in ways such that all team members would be able to come along for the ride; that job losses were an unacceptable cost of change. Encountering the queen of firefighting, who had spent a 30-year career in this mold, I came to realize that perhaps Jim Collins[1] had it right—you have to decide who you want on the bus. As a result, some people have to be asked to leave. Although leaders can set an expectation for a new culture and new social norms, not everyone will like it. Coaching may prove fruitless with such people, and meanwhile, their behavior may deteriorate. In 2007, the group project manager started to light fires just so people would see her putting them out. She became the organizational arsonist. She had to go!

1. http://www.jimcollins.com/article_topics/articles/good-to-great.html

Lessons Learned from Eli, #3:
Don't Assign Blame

Wednesday, June 30, 2004

• •

Remember, we get to choose the definition of "conformant quality." If we want to conform, we can always lower our standards.

• •

DON'T ASSIGN BLAME OR POINT FINGERS WHEN COMPLAINING ABOUT (AHEM, explaining) your delivery problems. It's all too easy to point the finger at someone elsewhere in the value chain and say, "I can't get my job done because _____ doesn't deliver _____ for me."

Eli Goldratt would prefer that we teach managers to express this in a less confrontational style, using the language of variation and conformant quality. "I can't deliver to expectations because my inputs suffer from [this] excessive common-cause variance and [these] specific special-cause variances."

What Goldratt is effectively asking us to do is to describe our challenges as system problems. It occurred to me that we could do this by plotting each element in the value chain on Wheeler's States of Control matrix. Hence, we might get explanations that sound like, "I can't deliver architecture predictably because my input, the requirements, exists in the Chaos state. We consistently rely on heroic effort from our team to ensure we deliver with conformant quality because of the unreliability of our inputs."

This concept asks us to define the notion of "conformant quality" at each step in the value chain. Remember, we get to choose the definition of "conformant quality." If we want to conform, we can always lower our standards. The Agile movement requires us to do this. The industry standard for "success," when used in the context of, "How many IT projects are successful?" is defined as "on time, on budget, with the required scope." The Agile movement argues that there is always too much uncertainty in the scope to bring it under control.*[1]

---

1. Authors such as Donald Reinertsen argue that attempts to eliminate uncertainty in the scope actually destroy value, as they discourage opportunity to take advantage of new information from the market and late-breaking demand from customers. Hence, the traditional view of conformant quality is ill conceived.

Hence, we should accept the notion of "on time and on budget, with most of the scope" as the new standard for conformant quality.

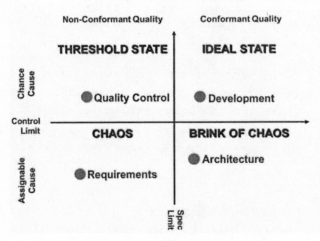

Figure 5.1 Donald Wheeler's States of Control matrix

How might it be possible to tighten up our definition of conformant quality and maintain that "on time, on budget, with required scope" definition? According to Eli's earlier lesson, we should reduce the batch size. With a smaller batch size, it is less likely that requirements will change during the processing time; therefore, the system will remain under control and continue delivering conformant quality. If we increase the batch size, the system moves out of control, forcing us to lower our standards.*

• • • • • • • • • • • • • • • • • •

* In the intervening seven years since this post was written, it has become evident that the chance-cause variation in our work is so great that, even with small batches, delivering conformant quality—where conformant is defined as "on time, on budget, with required scope"—is actually very difficult. Many teams practicing Agile methods such as Scrum report that they miss their commitments about 50 percent of the time. This lack of conformant quality occurs because of chance-cause variation in systems that are under control. This was a critical factor for me in developing the belief that a different approach was required. It was this belief that led to the pursuit of ideas that eventually emerged as the Kanban Method.

Conformant Quality
throughout the Workflow

Thursday, July 1, 2004

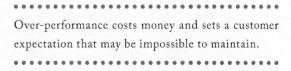

Over-performance costs money and sets a customer
expectation that may be impossible to maintain.

THERE IS AN ARGUMENT THAT SUGGESTS THAT THERE IS WASTE IF THE DEFINI-
tion of conformant quality is tighter or more exacting than the market (or customer) demands. Often, self-imposed definitions of conformant quality are
tighter than customers require. This tends to happen because core risk information from the source has been lost in the handoffs in the workflow, or because
managers along the value chain are measured in dysfunctional ways that encourage inappropriate behavior. Often, internal deadlines are created because someone believes they are needed to motivate people and to keep work flowing. It's a
management-by-objectives approach that doesn't even consider the possibility
of management by conformance (to process).

I have seen this with teams using Scrum, where they end up working at
heroic levels to achieve their sprint commitment (the definition of conformant
quality). In one conversation I observed, their customer did not actually require
them to deliver against a specific scope commitment, but they still felt the psy-
chological pressure to achieve it and were working overtime to do so.

Over-performance costs money, or has other sociological and physiological
costs that ultimately affect individual and team performance, and it risks setting customer expectations that may be impossible to maintain. Falling down
on previous expectations might erode trust. If this heroic behavior is invisible or
assumed to be normal, the customer will likely form an expectation of regular
delivery at that level of performance.

We must be prepared to rigorously determine what represents conformant
quality in the marketplace, and what represents acceptable tolerance in that
quality. These metrics give us the definition and position of the y-axis in our
Wheeler matrix. The tolerance in conformant quality defines the definition of

the columns. Shewhart used the term "specification limits,"[1] as opposed to "control limits" or "natural process limits," which determine the x axis and the row definitions on the Wheeler matrix.

Using this approach, and starting at the consumer, we can work our way back through the workflow to determine what represents conformant quality and tolerance at each step such that the consumer's demands are met appropriately.

What we achieve by doing this is a congruent alignment of functions in the workflow. Wheeler's matrix gives us a tool for understanding where change is needed and what its impact will be when it happens. Sometimes the change will happen in the market. The definition of conformant quality will move and challenge us. When external change happens, we will see each process in the value chain move from the Ideal state to the Threshold state (or the Chaos state if special causes appear as a result of our response to the market challenge). Each function manager must then work to bring their capability back to the Ideal state by implementing process changes and improvements.

---

1. Shewhart, *Statistical Method from the Viewpoint of Quality Control* (Dover Publications, 1986)

Lessons Learned from Eli, #4:
Lean and Six Sigma

Friday, July 2, 2004

● ●

I find that viewing a problem through just one of these lenses can often be limiting, or the solution is not obvious.

● ●

HOW DO LEAN AND SIX SIGMA INTERACT WITH THE THEORY OF CONSTRAINTS? It is a question asked often. According to Eli Goldratt, the Theory of Constraints provides a focusing mechanism for Lean and Six Sigma initiatives. In turn, Lean and Six Sigma provide the tools for exploiting the constraint (Step 2 in the Five Focusing Steps[1]) and subordinating to the exploitation decisions in step 2 (Step 3 in the Five Focusing Steps).

For example, when the constraint is in the market—that is, you can't sell enough of what you make—the exploitation requirement is that you achieve

1. excellent due date performance

2. short lead time

3. satisfactory quality (not excessive quality)

and do all three within the consumer's tolerance. What better mechanisms for this than Lean and Six Sigma, with their roots in understanding variation? Six Sigma's focus on reducing variation, and Lean's focus on eliminating waste and shortening lead times, will help improve throughput of the whole system by better exploiting the utilization of the capacity constrained resource in the workflow.

So, Lean, Six Sigma, and the Theory of Constraints are all complementary!

---

1. Goldratt, Eliyahu M. What is this thing called The Theory of Constraints and How should it be implemented? Great Barrington, MA: North River Press, 1999.

●●●●●●●●●●●●●●●●●●●

At the time of this writing (2012), I now view Lean, Six Sigma, and the Theory of Constraints as three different lenses for considering process improvements. Lean, with its focus on flow and flow time, Six Sigma, with its focus on understanding variation and reducing it, and the Theory of Constraints, with its focus on bottle-necks (constraints) and relieving them, are clearly complementary. I find that viewing a problem through just one of these lenses can often be limiting, or the solution is not obvious when understood using a single model. Multiple models encourage more innovative ideas for process improvement. I now teach all three approaches and I actively discourage dogmatic adherence to any one approach.

Naturally, each method has its own community, and communities, like any social group of humans, tend to be tribal in nature—and tribal membership is recognized by practices performed. I hope that the Kanban community we've created since 2007 has provided an umbrella that welcomes practitioners of all three approaches; it's ironic, really, that kanban—the word and the concept—is associated with Lean manufacturing, yet it has spawned an umbrella movement in knowledge work that provides a home for process-improvement thinkers from various schools of thought.

Socratic Method: Considered Dangerous When . . .

Sunday, May 2, 2004

• •

The person being questioned feels manipulated by the questioner . . . that it is all a trick; and most of all, he resents that the questioner already knows the answer.

• •

ELI GOLDRATT'S WRITING ON THE THEORY OF CONSTRAINTS CONTAINS A LOT of use of the Socratic Method—the style of teaching adopted by Socrates that questions pupils and asks them to arrive at the answers for themselves. Goldratt goes into some depth in his book, *What Is This Thing Called the Theory of Constraints and How Should It Be Implemented?* (North River Press, 1990. pp. 16–20), to explain why the Socratic Method works. However, I've learned the hard way that it is problematic, and more recently I've learned that Jonahs (qualified practitioners of TOC) all over the world have experienced the same issues and that the problem is well known and widely discussed. Well, evidently not well enough known. So I am documenting it for the benefit of those who come after me.

First, let's consider the problem statement. Generally, Goldratt recommends the Socratic Method when a systems thinker is trying to persuade a cause-effect thinker that a course of action is correct. The cause-effect person sees only immediate relationships and generally doesn't understand that in the big picture, things leverage each other within a bigger system. The purpose of the questioning, therefore, is to help this person connect a complex sequence of effects and understand how the system works. The questions are designed to reveal the layers of depth in the system and allow the cause-effect thinker to "see" the system and realize its deeply obscured influences, despite time delays and derivative actions between a cause and its effect.

Peter Senge called this point of linkage between cause and effect "the leverage point,"[1] and seeing it can be difficult. However, my experience with the

1. Senge, Peter M. *The Fifth Discipline: the Art and Practice of the Learning Organization* (Doubleday, 1994)

Socratic Method as taught by Goldratt has been very negative. In systems thinking and the patterns movement, they say you have to see something three times before you cease to think of it as coincidence. And I've seen this problem enough, now, to know that it is real and repeatable. So what is the problem?

Simply put, the person being questioned feels manipulated by the questioner. As layers of questions unveil layers of the system being analyzed, the person becomes increasingly irritated and objects to the questioning. He feels manipulated, as if the questioner is leading him down one path and somehow obscuring the other paths. He feels that it is all a trick; and most of all, he resents that the questioner already knows the answer.

When I first saw this happen, I put it down to American West Coast, liberal, conspiracy-theorist tendencies (I reside mostly in Seattle, Washington). However, I've recently heard that the problem occurs all over the world. At last week's Puget Sound TOC Learning Event, in Bellevue, Washington, coach Fran Fisher, who was an invited speaker, nailed the problem. Yes, the person being questioned feels manipulated, and the reason is simple: He knows that the questioner knows the answer, that the conversation is not one between peers. When the questioner is a knowledgeable expert in a field, and the subject is an apprentice, the relationship will break down if the apprentice feels manipulated. Fisher recommended that the method works if the relationship is a coaching relationship. In that case, the relationship is between peers. The person being coached should be treated as an expert in his or her field and the coach should be the novice asking innocent questions. The coach should know the right questions to ask but should not (generally speaking) know the answers in advance.* Indeed, this is truly a description of the relationship between Jonah and Alex Rogo in Eli Goldratt's novels. Jonah is not (or pretends very well not to be) an expert in Alex's field. He simply asks the right questions and forces Alex to come up with the answers for himself.

The revealing detail was finally offered from the back of the room, and I confess to forgetting who offered it. Apparently, Socrates used his method in two ways. The first was to ask a series of questions that led the pupil to a conclusion; then he asked more questions, which led to a conflicting conclusion. Socrates would expose the internal inconsistency and show people that, by their own logic, their proposal was invalid. He would also use the questioning method in collaboration with others when neither he nor his colleagues knew the answer. It was a journey of discovery for them all.

Hence, the advice seems to be, make careful use of the Socratic Method, as it is considered dangerous when the inquisitor is an expert and the inquisition is designed to educate the other party. It is perhaps better to use Socrates's method of questioning when no one knows the answer. Using it to flush out internal inconsistency risks embarrassing someone, so in that circumstance, it is best used only in private. Actually, I think that using the Socratic Method as a teaching tool is best avoided altogether.

● ● ● ● ● ● ● ● ● ● ● ● ● ● ● ● ●

* I've come to appreciate, in four years of consulting and training experience, that often the coach recognizes patterns that are common across firms in the same industry or, from time to time, simply common to all knowledge-work activities. As a result, the coach may well be able to predict the answers, but should not show this knowledge too early.

I'm often surprised by how surprised my clients are when, during a Kanban workshop, I can predict their problems or the nature of their work from the briefest descriptions. When asked to describe sources of dissatisfaction and identify types of work in their workflow, I am often able to interpolate many other issues they are facing from just a few data points.

The problem I identify in this article doesn't seem as prevalent with Kanban. I attribute this to the nature of the evolutionary approach to change. Kanban system design is about mapping current practice and exposing it in such a way that system effects are revealed for everyone to see, and in a way that connects with them emotionally as well as logically. In my opinion, a purely logical analysis, based on expert knowledge and experience and using Socratic questioning designed to reveal common system effects, is still likely to invoke a negative response

When Policies Move the Bottleneck

Monday, September 15, 2003

• •

When the bottleneck moves, everything changes. Changes in bottlenecks should be planned, and the overall system-wide effect understood in advance, otherwise chaos ensues, leading to accusations, blame, and loss of trust.

• •

INTRODUCING A NEW POLICY CAN MOVE THE BOTTLENECK, OR, AS IT IS MORE formally known, the capacity constrained resource (CCR), within a system. This can create problems throughout an entire organization if the effect of the policy changes is not understood. These problems are, however, foreseeable when a manager understands the effect of policies on the system. Goldratt initially referred to policies as a type of constraint. However, he later modified this advice to state that policies were only ever exploitation or subordination actions—constraints must, by definition, be capacity constrained, and a policy cannot in and of itself be capacity constrained. Introducing a new policy, however, might create a new system constraint, or bottleneck.

For example, let us imagine that the chief architect on a software project is becoming uncomfortable with the quality of emerging code and feels that the original architectural intent is not being achieved. With a simple announcement, he introduces a new policy that all designs must be reviewed by him before coding commences. In an instant, the chief architect is now the overall system bottleneck. A one-man bottleneck decreed by policy! It doesn't matter where the constraint was before—UI design, Testing, Usability, Requirements, or Coding—the new policy has moved the constraint.*

What are the effects of moving the constraint? People downstream in the process find themselves with idle periods. Work upstream begins to stockpile. People producing designs can't get to "complete" without a long wait. The project schedule can be abandoned; it, too, needs to be rewritten. The PMO (if there is one) needs to rewrite the master schedule. Customers get frustrated, and their expectations need to be managed. Why? Because the current plan was written to understand (implicitly) the current limitations of the system—the current constraint.**

When the bottleneck moves, everything changes.** New policies can move the bottleneck. Changes in the bottleneck need to be planned, and the overall system-wide effect understood in advance, otherwise chaos ensues, which leads to accusations, blame, and loss of trust. The introduction of new policies should not be made locally. The local benefit may be outweighed by negative system-level effects. New policies should be agreed upon system-wide after being fully understood by a higher-level manager or agreed upon by the set of line managers in the affected workflow.***

* * * * * * * * * * * * * * * *

* It's remarkable how many times this article has come back to haunt me in the intervening nine years. Policies that create one-man bottlenecks seem all too common. There seems no limit to the power of the human ego and the dissonance that goes with it.

* * * * * * * * * * * * * * * *

** When developing the Get Kanban board game simulation, Russell Healy chose to introduce a test manager character called Carlos. Carlos introduces a policy that creates a bottleneck in testing. Many people receiving Kanban training now experience and feel the effects of a policy-driven bottleneck firsthand. Ironically, it resonates with them all too often. It is my hope that, through training like this, we'll gradually see better quality-management decisions that better integrate negative system effects, such as bottlenecks, within the workflow.

* * * * * * * * * * * * * * * *

*** In 2003, I believed that a bottleneck in a software development workflow could be managed just like those identified in manufacturing processes. I've come to realize that common-cause variability is so great that it is rare (about ten percent of cases) that a bottleneck is clearly defined and remains consistent for long periods of time. In most cases, the bottleneck in the workflow can be moving around on a weekly—or even daily—basis. This is why kanban systems that limit WIP at each step in the flow work more reliably than trying to manage constraints using the Drum-Buffer-Rope technique from the Theory of Constraints.

Variability and Drum-Buffer-Rope*

Tuesday, April 27, 2004

● ●

Decrease the variability in the constraint while avoiding moving the constraint somewhere else in the system.

● ●

OVER THE LAST FEW EVENINGS, I'VE BEEN ACUTELY REMINDED OF HOW variability affects a Drum-Buffer-Rope flow solution for TOC. It's been pretty nice weather here in Seattle recently and the evenings have often been blue-skied and sunny. Every evening after dinner, I take my dog for a walk. In good weather, my daughter, Nicola, comes too. She is now 22 months old and very insistent that riding in the stroller is only for babies, and she prefers to walk. So she comes along, walking on her own. She also likes to hold the dog's leash. Now, Nicola weighs 25 pounds and the dog weighs 45 pounds. He is very strong, and he pulls on the leash a lot. He was, after all, half-bred to herd sheep down a mountain, with the other half designed to pull carts around China—definitely not the ideal combination to mix with a fragile toddler. Meanwhile, my daughter is still a novice at walking and she wobbles a lot. So, Dad has to be very careful and keep the dog on a tight leash, as illustrated in Figure 5.2.

Figure 5.2   Drum-Buffer-Rope system

As shown, I am acting as the Drum, setting the tempo, because my daughter, the Constraint, is too young and weak, compared to the pull from the dog, to act as her own Drum. The leash is the Rope in this example. The slack in the leash between Dad and Daughter is the Buffer. This is a Drum-Buffer-Rope system. Oftentimes, the buffer is twice the length of the rope between the dog, the input, and, Dad, the drum. However, it takes very little variation in the constraint—the rate at which a not-quite-two-yet little girl can walk—or variation in the strength of pull from the dog, the input, to cause the whole system to become unstable. When this happens, we all have to stop and reset. This makes our overall productivity, the rate at which we walk in the direction of the park, very low.

In this example, the rope is only a total of six feet long, and there is only two feet between the input and the drum. However, the desired production rate is perhaps two feet per second, or even more. The system is inherently unstable because there is insufficient rope and insufficient buffer compared to the desired production rate. The result is a two-year-old who is constantly in danger of being pulled over. In order to stabilize the system, either the production rate has to be lowered to a snail's pace—I have tried this and it works :-)— or the variability in the constraint needs to be greatly reduced—try telling that to a not-quite-two-yet toddler! An alternative solution would be a much longer rope to provide for a much greater buffer. With this in mind, I have acquired a 20-foot leash. This will allow me to provide up to 18 feet of buffer, which ought to be sufficient to facilitate a smooth walk to the park without incident or accident.

So there it is! With the Drum-Buffer-Rope solution, the production rate can increase in a stable fashion, only if you either (a) increase the size of the buffer or (b) decrease the variability in the constraint while avoiding moving the constraint to somewhere else in the system.

• • • • • • • • • • • • • • • •

* Over the years, this article has been widely acclaimed by many in the Theory of Constraints community as well as those coming to it as novices. Many have said that it can replace up to 50 pages of a complex textbook with a simple and accessible explanation that manages to include sufficient detail and nuance to make it useful to an advanced audience.

Typing the phrase "variability drum buffer rope" into Google.com, this article is the number one search result, some eight years after it was first published. It remains one of the most popular and widely referenced articles on agilemanagement.net. Strangely, this article represents perhaps 552 of the most useful words I've ever written.

Geeks and the Road Runner

Thursday and Friday, April 29–30, 2004

• •

Software developers are used to believing they are
the bottleneck. They like the idea that success lives
or dies with them.

• •

ROAD RUNNER BEHAVIOR IS THE TERM USED IN THE THEORY OF CONSTRAINTS
to refer to the ability to run at maximum speed for a period of time, then stop
and remain idle until new work comes along. Road Runner behavior is named
for the cartoon character, and it implies that local cycle times are optimized and
that slack is inherent in the system.

This behavior is common in some other lines of work; one that jumps to
mind is emergency services—firefighters, for example. When they get a call, they
must respond at the fastest safe speed. They are evaluated by how long it takes
them to respond to a call. Readers in the UK know that the government measures
the ambulance service this way, too. In fact, since the John Major government
in the 1990s, there have been standards for the acceptable number of minutes
in which to respond to a call. As a result, in slightly more remote parts of the
UK, where there isn't an ambulance base within a reasonable distance (in time),
extra ambulances are staffed and parked, often in the street of a small town, in
case of an emergency. This enables an ambulance to reach an emergency quickly,
and to provide paramedic care as soon as possible when the journey time from
the hospital or dispatch base would take too long. For emergency services, Road
Runner behavior is essential. The metric of the Road Runner is local cycle time.
No one counts the slack time.

So what do these firefighters and paramedics do when they aren't working (at Road Runner pace)? Do they sit around the firehouse and play cards,
as popular culture would have us believe? Do they goof off during those slack
periods? Or do they busy themselves with day-to-day maintenance, inspecting
fire hydrants, providing fire prevention consulting, and taking training?*

It might work for ambulance drivers, but by and large, geeks don't like Road
Runner behavior! Why?

I've heard this bleat a few times in recent months, "I've been under-utilized on this project"—interpretation: "I had slack time." Hmmm, let me think. Maybe you weren't the bottleneck. By definition, non-bottleneck workers should experience slack time. But geeks just don't like it.

The problem with slack is, I believe, two-fold. First, software developers are used to believing that they are the bottleneck. They like the idea that success lives or dies with them. If they work more, software is done earlier. If they work better, with higher quality, software is done earlier. If they work faster, software is done earlier. It's all about them and their capabilities. Secondly, they like bigger batch sizes because it allows them to work on their own without interference from bozo project managers and analysts for longer periods. In *Agile Management for Software Engineering,* I called this *Intellectual Efficiency.* They like to keep their minds busy, ideally with good, intellectually challenging work. They also like the freedom to decide when they need a break. Goofing off to play ping-pong isn't slack, it is recharging for more intensive intellectual work!

I personally hate task switching. My old brain moves slowly from one frame of reference to another, so I'd relish the down time. But it appears that I am not typical. The answer for geeks unlike me, who seem to be able to switch from one line of thought to another in a heartbeat, and who get immensely frustrated with the slack time in a Road Runner environment, is to give them planned bench projects or bench activities,** such as a reading or a learning program. Bench projects get done when they have a plan and are executed against that plan. The plan doesn't need to have a schedule, but it surely does need a feature list. A manager also can work with an individual to devise a training and a self-learning schedule. Slack time can be used for this, too. And it starts to pay back very quickly in improved quality and fresh ideas.

• • • • • • • • • • • • • • • • • •

* Readers familiar with Kanban will notice the observation that emergency-service workers have work in two classes of service—real emergencies, where the service requires Road Runner behavior, and other, non-emergency but important work that can be scheduled and performed at a less voracious pace.

• • • • • • • • • • • • • • • • • •

** In Kanban, this work is usually referred to as Intangible class of service work. It is important, but not urgent. Having such work as an integral and explicit part of the system actually improves its performance overall by reducing the likelihood of overproducing customer work and suffering the stockpiling, lengthening lead times, and the consequent costs of perishing knowledge that are typical of an unconstrained push system. Intangible class work helps to put idle slack time to good use and reduces or eliminates emotional resistance that might come from a fear of slack.

Improving a Constraint

March 11–12, 2005

● ●

It doesn't matter that we don't know where the
constraint is. We can still subordinate to it because
we know its capability.

● ●

Measuring Capability

I GET ASKED THIS QUESTION OFTEN . . .

How can you measure the capability of a team when every feature is different? My answer is always that capability isn't an absolute number (and I now realize that I didn't make this clear enough in my book, *Agile Management for Software Engineering*), and that it is expected to have a spread of values. It's important to understand the mean and the natural variation.

What's the first step in a Drum-Buffer-Rope solution for software development? Understanding the capability of the system.* Find a way of measuring it in terms that are meaningful to the outside—at the output, but preferably at the input. If the supplier—the marketing department, perhaps—provides Scenarios, the ideal unit of measure is Scenarios. If they supply Change Requests, then the ideal unit of measure is Change Requests.

To measure capability, simply observe the throughput of units of value over a period of time, for example, over a six-month period, sample the capability for each month. I recently worked with a team that does maintenance. Their unit of value is a Change Request. These vary immensely in scope, nature, and size. However, the manager collected data and demonstrated that there was such a thing as an "average Change Request," and that his capability could be defined in terms of a mean and a reasonable minimum and maximum per month.**

Subordinating to Capability

Step 3 of the Five Focusing Steps in the Theory of Constraints is "Subordinate everything else to the decision, in step 2, to exploit (and protect) the constraint, identified in step 1."

In the previous article, we identified the capability of our software development organization. We did this by measuring the output over time. We had no idea where the constraint was. Our system was opaque—just a mass of analysts, developers, architects, testers, and others doing work in some ad hoc fashion that we didn't attempt to understand. We didn't identify a constraint. But it doesn't matter that we don't know where the constraint is. We can still subordinate to it because we know its capability.

The average capability of the whole system, shown at the output, is the average capability of our constraint. And the spread of variation in the constraint must be no greater than the spread of variation we observe at the output. To subordinate appropriately in a Drum-Buffer-Rope problem, we must get agreement from the source to stem off the flow of input to the rate at which the constraint can consume it. Hence, if we measure capability as approximately 250 features per quarter, or 35 use cases, or 23 scenarios, this is the rate at which we should accept new work into the system. We need an agreement with marketing to supply only at the rate we can consume.

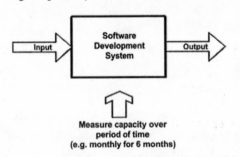

Figure 5.3 Measuring capability

What happens next?

Quickly after we stem off the flow of input to the rate the system can consume, our constraint is revealed. Why? Because non-constraints now have slack time! They have less work to do than previously, as we've replaced a push system with a pull system. Workers in non-bottleneck stations will start to put this slack time to good use—as I say, developers hate to be idle. When they are idle, they start to work on other things—like building tools, or environments, or building additional test environments, and so forth. The part of the system that is still capacity constrained continues to work flat-out on deliverables. This is how you identify the bottleneck in a system for software development or any other knowledge-work process that has a workflow:

❏ First, measure capability.

❏ Second, subordinate to that capability—and the constraint will reveal itself.

❏ Next, follow the Five Focusing Steps.

* * * * * * * * * * * * * * * *

* This article clearly shows us the emergence of what we now call the Kanban Method, particularly the Systems Thinking approach to introducing Kanban. March 2005 was clearly a pivotal point in my thinking and approach to managing software development. This represents the first significant moves away from the ideas in my first book, *Agile Management for Software Engineering*.

* * * * * * * * * * * * * * * *

** This manager was Dragos Dumitriu and the team was the now-well known XIT Sustaining Engineering team at Microsoft. At the time I wrote the article, they were almost six months into adoption of the first known kanban system for software development. The capability analysis I am referring to was performed by analyzing historical data from the software tracking system known as Product Studio, a forerunner of Microsoft's Team Foundation Server product. The approach they used is truly a prototype of the Kanban Method as we know it today.

Stop Estimating

March 13, 2005

• •

I've seen teams burn 40 percent of their capacity on estimating. That's 40 percent of their capacity wasted on something we know to be inaccurate at best and pure fantasy at worst.

• •

In this blog's tradition of cocking a snoot at traditional project-management guidance and conventional software engineering wisdom, I'd like to suggest that you STOP ESTIMATING. That's it—just say, "No!"

When I talked about subordinating to capability, I skipped over the second of the Five Focusing Steps—deciding how best to exploit the constraint. Exploiting (including protecting) a capacity constrained resource means that you do everything to ensure that its capacity is fully utilized. So why are you wasting that capacity estimating tasks or future work items—estimating future work that is at risk of obviation, or at the very least might require the analysis work done to provide an estimate, to be repeated when the actual work is queued for processing. I've seen teams burn 40 percent of their capacity on estimating. That's 40 percent of their capacity wasted on something we know to be inaccurate at best and pure fantasy at worst.

So I ask you, what do you think your customer would prefer? A promise that says, "We can deliver approximately 100 features this month, plus or minus 40" (that is, anywhere from 60 to 140), or a promise that says, "Well, we estimated everything carefully and we are confident that we can deliver 63 features this month" (. . .and we are almost completely certain that we will break this promise because this isn't an exact science)?* Actually, customers do prefer the false security of a precise promise. However, they definitely do not appreciate that promise being broken.

If you want to exploit your constraint fully—even if you don't know where it is—stop estimating! Starting planning based on throughput capability and embrace the concept of natural variation and fluctuation. Make promises based on approximate capability levels. Banish conformance to specification from your planning and estimating activity. When you can persuade everyone in the system to think differently, to view software engineering through a different lens, to embrace a paradigm of flow of value and variation, you can actually produce significantly more value for the customer.**

● ● ● ● ● ● ● ● ● ● ● ● ● ● ● ● ●

*The real issue is that customers really do prefer the latter. They love the certainty of it. It makes them feel comfortable. Your commitment and precision gives them reassurance. They are in denial of the latter clause that speaks the truth. Despite the recent evidence, they are in denial that you often break your promises. Writing in 2012, it's now evident that teams using accepted Agile estimating and planning techniques typically break their promises or customer commitments as often as 50% of the time. In random surveys of class attendees, I find that the range is 15% to 85% of iteration planning commitments are missed.

● ● ● ● ● ● ● ● ● ● ● ● ● ● ● ● ●

** In this article, I lay the groundwork for recommending two-point planning goals based on a study of the natural variation in throughput capability. I later embedded this advice into MSF for CMMI® Process Improvement, the method template supplied with Microsoft's Visual Studio Team System in 2005. I later discovered that the spreads of variation are so great, typically 1.4 to 2.0 times the value of the mean, that customers and other stakeholders are unwilling to accept plans based on such wide variation. As a result, I've abandoned the approach altogether. It is logically correct but psychologically unacceptable, and therefore not pragmatic or generally useful. Hence the approach described here has been replaced with the package of classes of service approach described in Kanban.

Drive Variability out of your Bottleneck

March 14, 2005

• •

If your bottleneck capability can fluctuate upward
significantly, it might cease to be the bottleneck.
When this happens, all bets are off. You have a new
bottleneck and you don't know where it is.

• •

YOU MUST DRIVE VARIATION OUT OF YOUR CONSTRAINT. I HARP ON THIS A LOT.
So do most in the TOC community. Why?

The most obvious reason, from this series of articles, is that it makes your planning more accurate. If you can say, "Our capability is 100 features, plus or minus 4, per month," that is a lot better than saying, "100 features, plus or minus 40."

The second reason is that wide variation means that your bottleneck might not be stable. If your bottleneck capability can fluctuate upward significantly, it might cease to be the bottleneck. When that happens, all bets are off. You have a new bottleneck somewhere else in your workflow, and you don't know where it is. If you have a different bottleneck, your plan based on throughput capability in a known bottleneck is no longer valid.****

• • • • • • • • • • • • • • • • • •

**** I have since come to realize that large amounts of chance-cause variation are normal in software development processes—and knowledge work generally. I have since abandoned the general advice to try and reduce variation sufficiently to guarantee that a bottleneck remains in a constant and predictable position. Unfortunately, dynamically shifting bottlenecks are a fact of life in the software development world. An entirely different approach is required This motivated me to adopt kanban systems. The realization that logical argument could not be used to overcome emotional objections led me to focus on organizational culture and encouraging the workforce to discover for themselves the changes that needed to be made. This is what we now know as the Kanban Method.

It is clear from this article that the Kanban Method can be described as an evolution of the Five Focusing Steps (Drum-Buffer-Rope) embodiment of the Theory of Constraints, most heavily associated with the 1980s. Goldratt's Thinking Processes emerged in the mid-1990s and are now widely regarded by that community as representing the main body of the method. Bill Dettmer, a well known author on the topic, once described the Five Focusing Steps and the Thinking Process to me as "two stores operating independently in a strip mall owned by one landlord."

In his 1990 book, *What is This Thing Called the Theory of Constraints and How Should It Be Implemented*, (North River Press, 1990) Goldratt identifies that the real constraint on change is emotional resistance from the workers and managers in the organization undertaking the change. As a reaction to this realization, he, with many collaborators, developed the Thinking Processes as a systemic approach to managing change. Coming to a similar conclusion about the problem with change, I chose not to pursue the Thinking Processes, and instead to develop an evolutionary approach. In purely evolutionary terms, we could think of the Thinking Process and the Kanban Method as evolving from a common ancestor.

Process Batch and Transfer Batch

Monday June 5, 2006

● ●

Iteration and release sizes will be different for every domain and, potentially, for every customer. Iteration length (and software development batch size) is a contextually specific problem.

● ●

I'VE FOUND IT USEFUL TO BORROW ANOTHER IDEA FROM THE THEORY OF Constraints and manufacturing planning to understand the dynamics behind a question such as, "How long should a project iteration be?" The concept has two elements: transfer batches and process batches.

A process batch is a batch of work that is done by a person, a team, or a system. Process batches are grouped for efficiency or for reasons of constraints, such as the size of a physical machine, or natural conditions, such as hours of daylight. For example, if you run a bakery, a process batch is the number of loaves of bread that you can bake in the oven in a single batch. Every batch has setup and a cleanup costs. Process batches tend to be optimized for efficient use of resources, efficient communication, minimum unit costs, or worker efficiency (based on minimum effort expended, which is measured as time-on-task or time-in-motion).

A transfer batch is one that is transferred down the value chain and passes to another set of workers or to the end customer. Transfer batches are often bigger, that is, several process batches make up a transfer batch. If you are in the bakery business, a transfer batch is the quantity of loaves you can load onto a truck to deliver to grocery stores. Transfer batches tend to be optimized for the costs incurred by the next stage in the value chain; that is, transaction costs incurred delivering to the customer, or costs incurred by the customer in taking delivery—for example, the customer needing to train a sales organization or a help-desk operation in the new functionality or features being delivered.

When I worked at Sprint, a large telephone company in the United States, a big concern for us was the cost of training the thousands of people in the retail stores and thousands more who answered the phone when customers pressed *2 on their phones for assistance.

Other aspects can come in to play. For example, do you have to take your web site down to do an upgrade? Will there be an outage? Is your business seasonal, so that you want to receive upgrades and new functions at only certain times of year? The latter concern means that the transaction costs vary at different times of year.

The transaction costs associated with a transfer batch can mean that transfer batches have to be significantly bigger than process batches—often many times bigger. In the Agile method, Feature Driven Development, the process batch is the Chief Programmer Work Package, and the transfer batch is a Release (or, sometimes, a Feature Set). Process batches are never larger than two weeks' work for a small team. Transfer batches are seldom smaller than three months' work. The customer often can't take delivery more frequently.

In much of the Agile literature* that talks about iteration size, there is no distinction between a process batch and a transfer batch. They are assumed to be the same. At the same time, there is no discussion about the transaction costs associated with a batch. We get advice such as, "Iterations should be one week (or two weeks, or four weeks, or even three weeks) in length." None of this comes with any consideration of the transaction costs associated with process efficiency or delivery (transfer) efficiency. The reality is that iteration and release sizes are different for every domain and, potentially, for every customer. Iteration length (and software development batch size) is a contextually specific problem.

* * * * * * * * * * * * * * * * * *

* I truly consider this thinking on iteration length, process and transfer batch size, and transaction costs to be an indicator that Agile methods can adapt and improve under the influence of Lean and Theory of Constraints. Six years later, advice on determining iteration lengths for Agile projects is still sketchy.

Reflections on the Passing of Eli Goldratt

Sunday June 19, 2011

• •

Eli never approved of an improvement that was achieved at the expense of the workers.

• •

ELI GOLDRATT, CREATOR OF THE THEORY OF CONSTRAINTS, AUTHOR OF MANY books, including the best-selling and seminal *The Goal,* passed away last weekend. I had the pleasure of meeting Eli twice and we interacted via email a few more times. Eli's work strongly influenced and inspired my own, and I'd like to share my thoughts on the sad occasion of his death.

I first encountered Eli's work when my boss at the time, John Yuzdepski, gave my colleagues and me copies of *The Goal* in March 2001. I read it on a flight from Dallas to Tokyo. About two-thirds of the way through, I had the epiphany— that I could equate features in Feature-Driven Development to inventory in Eli's Drum-Buffer-Rope system. I spent the remainder of the flight vigorously marking up notes in the margins throughout the book. What evolved from that was my 2003 book, *Agile Management for Software Engineering.*

Eli was kind enough to respond positively when he heard from me, out of the blue, asking him to review my manuscript and to write the foreword. He referred me to his friend and colleague, Eli Schragenheim, who provided the foreword. Goldratt was polite, patient, and welcoming to a newcomer to his community. I had not taken the classes, passed the exams, or become a "Jonah." I was just some guy who'd read seven books mostly written by him and who'd figured out some stuff for himself. I came to realize later that Goldratt's kindness extended to the professional courtesy that my use of his ideas was probably ill founded. He knew the underlying assumptions behind his own work, and his choice to focus on physical-goods businesses was not an accident. He almost certainly knew that the assumptions underlying his work were not true in knowledge-work problems like software development. He was kind enough not to shoot down my work, given that I'd spent a year writing a 300-page book about it.

I believe that Eli was incredibly tolerant of failure and of early innovation that wasn't fully baked or thoroughly thought through. We saw this with his tolerance of the early failures pitching "viable visions," and with his grace to hand off my manuscript to Schragenheim, who concluded that the mapping of requirements to inventory was "ingenious," and that it enabled a major step forward in managing software development. This was particularly poignant, as both he and Goldratt had owned and run software companies, and using Drum-Buffer-Rope had never occurred to them.

I was invited to the 2004 TOCICO conference in Miami, where I was to meet Eli for the first time. During the conference I learned that my own father had died. Eli was very warm and compassionate, which was surprising to me, as I was a total stranger to him. I was unable to get a flight back to Seattle to collect my passport, so I actually stayed through the end of the conference, returning home on my scheduled flight, only to turn around and fly to Glasgow immediately afterward.

The following year, I was invited again to TOCICO, where I presented the XIT case study as a DBR implementation. This was the foundation of what we now refer to as Kanban for software development, and it had evolved directly from my experience writing my first book. I was inspired by Eli's Five Focusing Steps and the incremental approach to change inherent in TOC. Once again, I found Eli warm and welcoming to an outsider who was presenting ideas and experience from a different field. It was evident that he was fiercely loyal to his people, as well as affectionate in a fatherly sort of way. Everything he gave he got back tenfold from his loyal following.

I will remember Eli Goldratt primarily as a warm, affectionate, kind, loyal, trusting, and respectful man who just happened to care deeply about how people worked together. Although he held a passion for effective performance in business, Eli never approved of an improvement that was achieved at the expense of the workers. One might consider him a highly social capitalist. I believe his leadership in this respect is exemplary, and that in the future, it will be recognized as visionary and ahead of its time.

I have the greatest respect for Eli, the man and the intellectual, and his work continues to inspire me, even as my focus has changed and I've come to see the limitations of applying TOC to knowledge-work problems. To recognize limitations isn't to criticize. Today we're solving a different set of problems from the ones Eli tackled. Eli's work will continue as a core foundation of how we think about improving effectiveness in the knowledge workplace. Eli's leadership and example will continue to inspire us as we seek the intersection of performance that brings a positive economic outcome together with a better sociological outcome for everyone involved.

Eli Goldratt, may you rest in peace. My deepest condolences to your family and everyone at Goldratt Consulting.

Some Holiday Constraints

Saturday, December 27, 2003

I WAS WATCHING THE DVD OF *RUDOLPH THE RED NOSED REINDEER* with my daughter last night. I was amused to note that the constraint in the elf toy factory is the paint shop operated by an elf called Hermie—I was remembering the "walk in the hills" example from *The Goal*, where the slowest boy is Herbie. How do we know Hermie is the constraint? Because the foreman sees the inventory building up in front of him. Why is Hermie the constraint? Because he is not motivated in his job; lack of proper motivation reduces his throughput. Hermie wants to give up being the factory constraint and become a dentist instead. As I've pointed out in the past,[1] dentists really understand how to be the constraint and how to manage around it.

I see a parable in the life of Hermie and the elf toy factory. I'm not the first person to observe that software engineers need to be properly motivated in order to be productive, and I have often talked about the role of leadership and management in creating a properly motivating environment for software engineers. I may have been the first person, however, to point out that lack of motivation can cause developers to become the system constraint. I spend a lot of effort in my book, *Agile Management for Software Engineering*, making this clear, and I give techniques to help a manager identify the constraint. In the case of a lack of motivation, I delegate to others the problem of elevating the constraint. There has been much written by the likes of Jerry Weinberg, Tom De Marco, Tim Lister, and Larry Constantine on how to motivate software engineers. I don't need to repeat their advice.

1. http://www.agilemanagement.net/index.php/blog/Throughput_and_Americas_Dentists/

In the movie, the foreman elf "motivates" Hermie by telling him that he should put up and shut up, and that his life is that of a toy factory elf—forget becoming a dentist! The result is that Hermie drops out and becomes a "misfit." I'm sad to reflect that I have seen many similar attempts at motivation over the last two years, during the recession in IT. All too often, management tells developers that they are "lucky to have a job," and to "stop asking for help," and "just get busy coding."

My prediction, then, for 2004, is that as the economy improves (even if this is temporary), it will unleash a backlog of pent-up frustration, and we will see excessive staff churn as geeks move on in search of better bosses; better, more motivating environments; and new challenges.

On Constraints and Transportation Systems

CHAPTER 6

T HERE ARE MANY REAL-WORLD EXAMPLES OF FLOW. WE STAND IN LINE TO BUY THINGS. WE TRAVEL FROM PLACE TO PLACE. WE ORDER THINGS TO BE DELIVERED TO US. WHEN FLOW IS INTERRUPTED OR UNEVEN, WE FEEL IT AS DISSATISFACTION AGAINST OUR EXPECTATIONS. BUSINESSES experience it as excessive cost and dissatisfied customers.

The same levers we use to manage the flow of work through a development workflow are available to us in real-world situations. We can adjust batch size, quantity of work-in-progress, location of bottlenecks, and shape the variability we encounter. The mechanisms available to us to control these levers are management policies.

In Agile methods, during an iteration timebox, the team is protected from outside interruptions; at the same time, daily communication within the team intensifies. This should limit the introduction of avoidable external variability (interruptions), while quickly addressing project-related variability (issues). During a retrospective, observations can be made about flow; that feedback can be used to make policy changes for the future. All of these practices and policies contribute to a smoother, steadier flow.

Kanban asks us to model existing workflow and express it with a visual model. The design of the visualization might allow us to zoom in on particular areas of interest within the workflow, which enables us to manage them better.

Flow problems are all around us in the physical world. It is useful to think about flow when you are standing in line to order coffee or waiting to take the train to work. Note the elements that affect flow for better or worse. Can changes happen that would bring the win-win of an improved customer experience and an improved business value? Kanban enables us to visualize invisible work and invisible workflows. By doing so, it helps us engage all our senses and use our emotional intelligence as well as our logical capabilities. What we learn from physical environments, such as cafes, airports, trains, buses, and even cable cars, can then be applied to improve our working processes in our office full of knowledge workers.

Thoughts on the
Retirement of Concorde

October 27, 2003

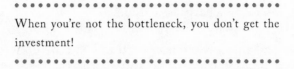

When you're not the bottleneck, you don't get the investment!

IT WAS WITH SOME SADNESS THAT I FOLLOWED THE NEWS OF CONCORDE'S LAST commercial flight. I remember, as a small boy in the early 1970s, going to Prestwick International Airport to see the prototype. It is one of my earliest memories. Concorde isn't an aircraft—it is an icon. Concorde is the quintessential essence of London, the crown jewel of the national flag carrier, British Airways. You cannot arrive at London's Heathrow Airport without seeing the 1:5 scale model. You cannot avoid the souvenirs of London and Britain that carry its image. Apparently the French had something to do with it, but no one really remembers that anymore. Its wings and engines are British—what else is left? The drooping nose, perhaps? The main contribution from the French—in what is perhaps the archetypal instance of industrial espionage in the late twentieth century—was to leak the blueprints to the Russians, which resulted in the TU-144.

In the days when America was putting men on the moon, the world's only supersonic passenger aircraft was British, and that was something small boys could be proud of—something inspiring. That's what makes Concorde an icon of Britishness in the latter twentieth century.

So why is it gone and without a replacement? Why are slow, fat, large, ugly aircraft all the rage?

Simply put—for most passengers, the vast majority in fact—time is not the constraint. It makes more sense to go slowly and save money. For businesses, this reduces costs. Reduced costs go straight to the bottom line. Greater speed would be worth it only if it increased throughput. This would be true only when the passenger is a constraint (or works within one). Most businesses just blindly try to save costs on corporate travel. Few businesses have an idea where their constraints on increased throughput lie. The result is that saving costs blindly does, ironically, make sense. The chance of impact on the constraint is slim.

However, when a constraint's capacity is reduced through slow travel, there is a loss in throughput for the business.

Why is this not much of an issue in the early twenty-first century? As the IBM advertisement[1] used to say, "Show me the flying cars! They promised us flying cars!" and then explained that we don't need them because we have the Internet. So there is no need for a supersonic jet airliner in the twenty-first century, then? I think that is a hasty conclusion.

Some market research, TOC-style: Who would want to justify the costs of flying supersonic? Someone who is a constraint on throughput? Who are such people? Celebrities, for the most part. There is only one Ewan MacGregor, for example. If you hired Ewan MacGregor, only the physical presence of the man himself will suffice. There is no substitute available. If a celebrity, such as Ewan MacGregor, wants to maximize his revenue-generating potential, he needs to get places fast. So it is no surprise, then, that before the tragic accident, Concorde was very popular with the showbiz glitterati.

So, is there a market for a replacement? If a plane could be built cheaply enough, and a limited number were made for routes such as New York–London, Los Angeles–London, Los Angeles–Tokyo, it just might be possible. The *Economist* thinks that there are alternatives, such as private jets[2]—more flexible substitute products. They are not as fast in the air, but the total door-to-door time is comparable with Concorde.

And therein lies the lesson—the time spent in the air is not the system constraint! The goal for a celebrity is door-to-door transport in the shortest possible time. Spending billions to shorten the time in the air might not be the best use of funds. There might be better return on investment in reducing the time spent getting to, from, and through airports, and providing flexible flights in small, private jets from quiet, private airfields.

Britain continues to lead the world in fast cars—almost all racing cars are made in Britain, even the ones that go around in circles in the United States—and Britain iconified the fast passenger aircraft. Why couldn't it brand itself "fast" Britannia, and niche-market its products as fast, good, and expensive? Britain needs something to be proud of. The Millennium Dome[3] just didn't hack it. A new Concorde project might be the thing—something around which people could rally, something to be proud of, something to print on all the tourist souvenirs for a new century. However, the economics just don't seem to be

1. http://slate.msn.com/id/1005883/

2. http://www.economist.com/PrinterFriendly.cfm?Story_ID=2142593

3. http://www.economist.com/PrinterFriendly.cfm?Story_ID=2142593 rope/07/26/concorde.photo/

there. Building a new Concorde appears to be a poor use of funds, even when the derivative effects—inspired little boys who grow up to be productive in other ways—are fully taken into account.

Bottom line—when you're not the bottleneck, you don't get the investment. Sorry, Concorde! I will miss you.

Speed is the Essence of Value Delivery

September 30, 1999

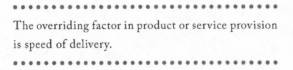

The overriding factor in product or service provision
is speed of delivery.

DURING THE PAST FIVE MONTHS, I HAVE BEEN WORKING IN IRELAND ON
weekdays and living in Scotland on the weekend. My weekly commute between
my home on the west coast of Scotland and my apartment in Ballsbridge, Dublin,
takes just over two hours, at best, and three-and-a-half hours on a bad day. I take
a half-hour flight from unfashionable Prestwick International Airport to Dublin
with the basic, no-frills Ryan Air.

There are several travel alternatives for me. There are numerous ferries from
Scotland to Ireland. I could, for example, journey from Troon to Belfast, and
then on by car or train to Dublin. The total time would be around five hours on
a good day. Rather slow, and the cost is perhaps slightly more than flying.

The most obvious alternative would be to fly from the much larger Glasgow
Airport to Dublin using the Irish national carrier Aer Lingus—a traditional airline that offers a high standard of service and modern aircraft. Glasgow Airport
is equidistant from my house to Prestwick. So why do I choose to fly from sad,
somewhat dilapidated (but improving) Prestwick using low-end Ryan Air, which
offers a no-frills service in very, very old aircraft?

The answer is simple—speed!

The Journey

It is easily the fastest route. My goal is to get to my destination. My task is to
travel. My aim is to achieve the goal in the fastest time with the minimum stress.
Only Prestwick to Dublin offers this. How?

Prestwick airport is situated at the end of the A78 trunk road. To get there I
turn left out of my driveway and proceed down the A78 for about 20 miles. Most
of the route is Motorway class road (for an American audience, this equates to
a four-lane freeway with on and off ramps, overpasses at intersections, and no
traffic lights). I can get there in about 20 minutes.

Once there, the parking lot is close to both the main road and the terminal building. Prestwick is a small airport that handles a small quantity of traffic. Parking is easy. Walking to the terminal is quick and easily navigated.

Checking in is a pleasure. There is often no queue. Check-in can be as late as 20 minutes before takeoff.

The flight is quick, only 30 minutes, and usually it runs on time, often arriving early.

At the other end, Dublin is busier and more congested, but it is still a short walk to the rental car desk, and a short walk from there to the pickup point.

The final leg, driving through Dublin, is a chore, but on a quiet Sunday evening it's no problem. I am home only two hours after leaving home at the other end.

Alternatively, I could go to Glasgow on a busy, narrow, winding road, which is often packed with commuter traffic or slow-moving heavy vehicles and farm machinery. The parking lots are vast and crowded, finding a space can be difficult, the walk to the terminal is long, and the check-in queues can be long at peak times. In comparison, though, Glasgow is a feature-rich airport with good shopping and a nice range of cafes, restaurants, and other services.

The Lesson

The lesson here is that the overriding factor in product or service provision is speed of delivery. The journey is stressful. Reducing the time and easing the friction along the way is important.

My goal is to get to the destination. I do this by making a journey. There are many factors involved: road to the airport, ease of parking, journey time from parking lot to terminal, check-in process, airline, aircraft, flight time, ease of rental car check-in, journey time to pick up the rental car, and road from the airport at the other end.

This provides an interesting example, as no one provider for air travel can control my decision. Aircraft designers can design a nice airplane, but that won't make me use it. An architect could design a nice airport, but that won't make me use it. Fast roads provided by the local authority won't make me use them, either. Nice service and an appropriate flight time from the airline won't make me use it. It is the whole chain of services related to achieving my goal that affects my decision. The "whole product" that delivers my goal is a road network, an airport, an airline, an aircraft, a rental car company, and then another road network.

Delivering Door-to-door Value

October 28, 2003

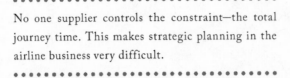

No one supplier controls the constraint—the total journey time. This makes strategic planning in the airline business very difficult.

MY POST ABOUT CONCORDE PROMPTED A DISCUSSION THAT CAUSED ME TO recall my 1999 article about speed of delivery. In it I discuss that in the airline industry, true business value should be delivered as an end-to-end solution—the fastest time from door to door.

> My goal is to get to the destination. I do this by making a journey. There are many factors involved: road to the airport, ease of parking, journey time from parking lot to terminal, check-in process, airline, aircraft, flight time, ease of rental car check-in, journey time to pick up the rental car, road from the airport at the other end.

What is really interesting about the conclusion in that article is that no one supplier controls the constraint—the total journey time. This makes strategic planning in the airline business very difficult. Is the answer vertical integration—control the value chain, control the constraint—or collaborative, open, transparent, partnership? The company (or companies) that work(s) this out will conquer the travel business.

David Taylor describes this dilemma as "Winning as a Team" in Chapter 3 of *Supply Chains* (Addison Wesley, 2003). He points to the trend toward partnership between companies with a core competence at one point in the chain rather than vertical integration. He calls well-formed partnerships "virtually integrated" businesses. He points out, however, that truly transparent, collaborative partnerships might be an unobtainable nirvana. There is a basic conflict between such mutually agreed-upon partnerships and capitalism. The need to improve profits ultimately makes companies want to squeeze their supply-chain partners.

If some group of businesses does work this out, they won't be the only winners. We, the traveling public, will win too. Business can be really simple—align your business goals and your ability to make profits with the goals of the customer.

Concorde and Six Sigma

Wednesday November 5, 2003

●●●●●●●●●●●●●●●●●●●●●●●●●●●●●●●●●●●●

First, a business must face brutal reality by believing
its metrics; secondly, it must be gathering the correct
metrics.

●●●●●●●●●●●●●●●●●●●●●●●●●●●●●●●●●●●●

FIGURE 6.1 SHOWS CONCORDE TAXIING INTO ITS RETIREMENT HOME AT BOEING
Field in Seattle, Washington, at three o'clock this afternoon. As I wrote in
Thoughts on the Retirement of Concorde (page 139), I was lucky enough to see
its first incarnation, and now I was able to witness the arrival of the last one on
its last official flight. For good measure, it set the passenger-jet speed record
for an East-to-West-coast crossing of North America—thanks to the Canadian
government, which let it fly supersonic.

Figure 6.1   Concorde in retirement

So why is Concorde sitting here on the tarmac in Seattle, a museum piece,
rather than flying? It is, after all, a young air frame. With proper maintenance,
Concorde could have kept flying for some time to come. On July 25, 2000 a
Concorde exploded just after takeoff at Charles de Gaulle airport outside Paris.
After some extensive modifications, the fleet re-entered service, but consumer
confidence had dropped, and its mega-rich customers didn't come back.

As mentioned previously, the niche market for Concorde was passengers who
are their very own capacity-constrained resources—for them, time is money (or
some other goal, such as, time is leisure). For the mega-rich who needed more time
to shop on Fifth Avenue, Concorde was the only ticket to have. Those customers
didn't come back. Why? Joan Magretta, writing in *What Management Is* (Free
Press, 2002), uses two key ideas in management science to explain why Concorde
was grounded. First, a business must face brutal reality by believing its metrics;
secondly, it must be gathering the correct metrics. The Six Sigma principles of

measuring for quality of four failures per million opportunities is important with safety-critical systems in which the customer must have absolute faith.

The numbers that matter are the ones that help you to face reality, and to do something about it. Consider:

> On July 25, 2000, an Air France Concorde jet exploded shortly after take-off . . . The crash left 113 people dead. These were the first Concorde fatalities in the supersonic jet's thirty-one-year history. Within days, the Concorde . . . was grounded indefinitely.
>
> Why was the world's fastest passenger jet suddenly taken out of service? There had been only three disintegrating [tire] incidents over thirty-one years, and none had caused a plane to crash before. How bad a safety record is that?
>
> In a word, unacceptable—but that is clear only when you compare the number of incidents to the total number of times the Concorde has flown. Because there were so few jets in service and so few flights per day, the failure rate was extraordinarily high. If that rate were applied to a fleet of U.S. airlines in service, it would produce one serious [tire] explosion . . . per day. One per day! That is why Concorde was grounded. And it is why management requires the discipline of quantification. Simple numbers help us to face reality and to make sense of the events in ways that our intuition alone cannot do.[1]

Why the customers didn't come back is an area for psychologists. We humans are very poor at risk assessment. We tend to overestimate some risks and underestimate others. We try to scale risk taking against the consequences. In this case, the perceived risk of death from flying Concorde was very high—and rightly so, as Magretta points out. British Airways didn't do enough to convince the customer base—that elite niche of Concorde passengers—that the problems had been dealt with appropriately and completely. They failed to communicate the safety improvements. The customers failed to believe and showed, through lack of patronage, that they valued their lives more than the time saved on transatlantic crossings. Despite the fact that such people perhaps take more risks with their lives through other lifestyle choices, for them Concorde seemed like a risk too far. Where human risk assessment is concerned, perception is everything.

In recent months, British Airways filled the jet with "supersonic virgins," those willing to risk their lives for the flight of a lifetime. With no solid customer base to sustain it as the flagship service, its value to the BA brand was diminished, and consequently, its fate was sealed.

---

1. Joan Magretta, *What Management Is*, pages 119-120, Free Press, 2002

Seattle's Free Buses:
A Constraint-Centric Explanation

Sunday, September 7, 2003

• •

Often the throughput, or effectiveness, of a system element can be limited by a policy constraint. The policy might be having a negative effect on the whole system.

• •

SEATTLE HAS A RIDE-FREE[1] BUS SERVICE WITHIN A LIMITED ZONE AROUND downtown. It operates from 6:00 a.m. until 7:00 p.m. every day. People have speculated that it is a county government perk for tourists, or perhaps for office workers, or maybe the politicians just want to be nice to the voters! I think this is unlikely. Perhaps systems thinking and the Theory of Constraints offer a better explanation.

Public transport systems can develop virtuous or vicious cycles—the more they get used, the more service is provided; and the more available the service, the more usage. Equally, the corollary is true: the less it's used, the less service is provided—which leads to less usage. There is a subtle tipping point between the two, and a stable system seems hard to achieve. If service is unreliable and infrequent, the paying public will go elsewhere and usage will fall. This will lead the supplier to reduce service, and soon there is little benefit at all from a public transport system.

So it is desirable to have both timely and frequent service. If buses are to run on time, uncertainty surrounding the schedule and the timetable must be reduced. Handling large numbers of passengers at a few stops downtown has more irregularity than handling smaller numbers at many stops in the urban neighborhoods and suburbs. However, if passengers don't have to mess around finding change and drivers don't have to collect fares, the irregularity associated with large numbers of people entering or leaving a bus is reduced. Making the downtown area a ride-free zone improves the likelihood that the timetable can

---

1. http://transit.metrokc.gov/tops/bus/ridefree.html

be met. It helps to keep buses moving and to keep downtown traffic flowing. This systemic effect encourages usage through improved quality of service.

All that was required to make this happen was a change of policies—the policies that state the rider must always pay and that payment must always be made on entry. As soon as these policies were waived in favor of one in which inbound riders pay on entry and outbound riders pay on exit, irregular flow was reduced and buses were more likely to run on time. The side effect—and added passenger benefit—was the emergence of a ride-free zone. The transit company trades off the localized revenue lost downtown against the gain from increased ridership by longer-distance travelers.

Often, the throughput, or effectiveness, of a system element can be limited by a policy constraint. The policy might be having a negative effect on the whole system. Eliminating the policy might produce both a local improvement and a global improvement—as is the case with the King County Metro Ride-Free Zone—buses run on time, but even better, the bus service is used more overall.

Back on the Buses

Thursday, January 29, 2004

• •

A big driver of variance is how many people get on
or off and how quickly they can pay the fare. Elderly
and disabled people increase the variance on stops.

• •

As I mentioned in Seattle's Free Buses (page 147), I commute to
downtown Seattle using the King Country Metro bus service. Over the New
Year's holiday, KC Metro took delivery of some new rolling stock. The new
coaches can be distinguished from the older ones because they have a low floor
with only a single step into the main cabin. The low floor is achieved by moving
the engine to the back of the bus, where it is mounted vertically. There is no rear
window in a low-floor bus. In addition, the front wheel wells protrude into the
passenger cabin, which reduces seating capacity.

So here is the Theory of Constraints explanation for why low-floor buses
make sense most of the time, but why they don't always make sense on my oft-
traveled routes—numbers 18, 15, and 17.

As noted previously, the usage of a public transport system depends on the
frequency of service and the reliability of that service to run on time. With a
frequent, punctual service, a virtuous cycle develops—travelers use the service
more and come to depend on it more. As soon as either frequency or punctual-
ity drops off, passengers go looking for their car keys, and a vicious cycle has
started. So systems thinking and constraints thinking go hand in hand to provide an efficient public transport system.

Now for the constraint thinking: The new, low-floor buses provide ease of
access for those with mobility challenges, such as elderly people. It is easier to
get on and off a low-floor bus. It is also faster for such people to get on and off
when there are fewer steps. Those in wheelchairs or who use a walker require
the use of a lift. A low floor, coupled with a self-lowering suspension, allows the
lift to extend faster. In fact, I timed it as 30 seconds faster to load a passenger
on a low floor bus versus a standard one. Again, timeliness is important to the
service. The bus will suffer variance in its schedule as it travels through traffic.
A big influence on variance is how many people get on or off and how quickly
they can pay the fare. Elderly and disabled people increase the variance on

stops—particularly when the lift is used. Providing low-floor buses is a win-win. It gives better quality of service for those who need the benefits of a low floor and it helps reduce variance, which helps to keep the buses running on time.

So low-floor buses are a boon, right?

As we know, the low-floor design reduces seating capacity. In fact, at least eight seats are lost to the intruding wheel wells and the vertically mounted engine. This is not a problem when capacity is not the constraint. When the constraint is purely adherence to the timetable, and the bus is underutilized, eight fewer seats is never a problem. However, because route numbers 15 and 18, and to a lesser extent, 17, are the most heavily used routes in the city, they were chosen to be replaced by the future monorail.[1] During rush hour, the 15 and 18 buses are capacity constrained. It is common for the standing room to be full and for passengers to be turned away. The introduction of the low-floored coaches has only exacerbated this problem.

How can this situation be improved? Clearly, reducing quality of service for elderly and disabled passengers is not the answer. The truth is that the monorail replacement cannot come soon enough. The monorail will provide more capacity, and by elevating (literally) the problem above the traffic, variation will be reduced and challenges with adherence to the timetable will no longer be a concern. Roll on, December 2007!

● ● ● ● ● ● ● ● ● ● ● ● ● ● ● ● ●

*Subsequent to writing this article, a campaign to torpedo the monorail project as too expensive took hold. In a plebiscite Seattle, voters narrowly defeated the project. Ballard residents continue to commute into the city center via the numbers 15, 17, and 18 buses.

1. http://www.elevated.org/

On the Campus Buses

Wednesday, March 2, 2005

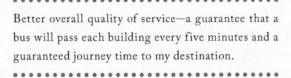

Better overall quality of service—a guarantee that a
bus will pass each building every five minutes and a
guaranteed journey time to my destination.

I USE PUBLIC TRANSPORT A LOT. IN FACT, I OFTEN GET UP AS EARLY AS 5:15 A.M.
to catch a bus to arrive at the Microsoft campus around 6:30. Such is life when
you want the convenience of coming home in the high-occupancy vehicle (HOV)
lane* but still have to be there for a 7:00 a.m. meeting. Once at the Microsoft
campus, I can take a minibus from the Overlake Transit Center to any Microsoft
building. The service is convenient, fast, clean, has excess capacity, and there is
free candy—you don't get that on King County Metro!

Back in September, Microsoft made some significant changes to the on-
campus bus service. Until then, the service had been door to door, like a cab
service. You simply asked a receptionist to call for the bus and you waited a few
minutes. It would come to your building and then take you directly to your destination. Occasionally, someone else might be using it and you'd have a minor
diversion to stop at another building. The whole system relied on a dispatcher
to direct the buses for optimal flow. In summer, the service was great—pickup
within ten minutes and delivery to destination within ten minutes. However,
in winter, when demand was greater, the wait could be up to 45 minutes. The
system needed to change.

The company introduced a circular, timetabled route—in fact, two routes,
a clockwise and a counter-clockwise route. Buses run at ten-minute intervals
staggered five minutes in either direction. Buses don't stop at every building, but
at designated stops situated centrally within a cluster of buildings.

When I first heard of the changes, it set off my "cost accounting" sensor
and my nose started to twitch. I used the bus service pretty much every day, so I
started talking to the drivers. I was actually on one of the last-ever door-to-door
service vehicles just before 5:00 p.m. on the changeover day. And I was one of
the first passengers on the scheduled bus when it started the following Monday
morning. So I got to experience the before and after quite closely together. I was
expecting to discover that management was secretly calculating efficiencies for

trips, drivers, and buses, and that the changes would "increase efficiency." I expected that the new fixed-route service would look more efficient but that there would actually be no true savings because the costs were fixed—same number of vehicles, same number of staff. As it happens, I was surprised. The new service is a win-win.

The new, fixed route provides better overall quality of service—a guarantee that a bus will pass each building every five minutes and a guaranteed journey time to each destination. In winter, this is a huge improvement over the old door-to-door service. So the customer—the traveling employee—wins. The customer throughput, on average, is increased. It also turns out that the drivers are hourly paid variable costs; sure enough, the fixed route allowed management to reduce the number of buses actually working and cut driver hours as a result—a true cost saving. Note: this is a classic Lean strategy of smoothing flow and reducing WIP inventory. The airlines are doing similar things in their hub airports, smoothing the flow of arrivals and departures and moving away from "waves" of flights. This has enabled them to reduce the number of staff required. So there is a throughput accounting gain for the business, too. It's a genuine improvement, not just a cost-accounting mirage. This is a good example of a cost-saving initiative done right.

● ● ● ● ● ● ● ● ● ● ● ● ● ● ● ● ● ●

* During the evening rush hour, the Washington State Route 520 is often backed up for several miles. I described this problem in depth in Kanban: Successful Evolutionary Change for Your Technology Business. Buses are allowed to use the HOV lane and avoid much of the jam. This can save as much as 30 to 45 minutes of journey time. Hence, riding the bus to work in the morning, rather than driving my car, is necessary in order to enjoy the HOV lane option in the evening.

Bypassing Hubs: Where is the Constraint
in the Long-haul Airline Industry?

Monday, October 27, 2007

• •

Boeing's approach is Lean—it is based on smaller
batch transfers with smaller aircraft. It means
nonstop flights to and from smaller, less crowded,
traditionally secondary airports, or minor hubs.

• •

THE *Economist* HAS RAISED THE LEVEL OF DEBATE IN THE ECONOMIC WAR
between the two great manufacturers of long-haul aircraft—Boeing and Airbus.
As a Seattle resident with an interest in the local economy, I care deeply about
the success of Boeing. So who has got it right, Boeing or Airbus?

While the locals here in Seattle can take some pride in the fact that Boeing
appears to have pulled one over on rival Airbus—its 787 Dreamliner model
seems to be generating many advance orders—do they have the strategy right?
Presently, Boeing enjoys several advantages over its rival. Its lean manufacturing
capabilities—huge aircraft assembled on a moving line as well as greatly reduced
WIP inventory and lead times—are reducing costs compared with Airbus. The
strong euro means that Boeing's planes are more competitively priced. It looks
like Boeing stands to be the big winner in commercial jet production in the
next decade. The article in the *Economist* gets at something deeper, though—the
guesswork required to predict the future constraints on the long-haul airline
business and how best to exploit them.

Boeing believes that a combination of airline deregulation and
the popularity of heavy-twin aircraft have changed long-haul flying for
good. Instead of the hub-and-spoke system, in which passengers flew in
747s to big hub airports and then took short-haul flights to their final
destination, Boeing says that passengers now want the convenience of
flying point-to-point and that smaller long-haul planes make it both
possible and economical for them to do so. As evidence, Boeing points to
the drying-up of orders for passenger versions of the 747.[1]

---

1. "The Giant on the Runway," The *Economist (need date and page numbers)*

The article then debates the cost-accounting efficiency model that clearly guides the strategic planning at Airbus. It also debates the point-to-point heavy twins versus the hub-and-spoke model that started with the 747 and that Airbus believes can be further enhanced with the A380.

If Airbus is to win, the constraints have to be air-traffic control systems limiting expansion at more airports, environmental concerns, physical constraints limiting growth of runways and takeoff slots at major hubs, and perhaps the number of qualified pilots. Airbus would exploit the constraint by simply flying a bigger aircraft—the A380. This allows more passengers to pass through the bottleneck. The exploitation mechanism is a bigger batch transfer. It allows more people to fly without more runways or takeoff slots. It is also efficient from a cost-accounting perspective and drives a lower cost per passenger mile.

Those of us who've been paying attention to the Lean revolution that is quietly dominating businesses well beyond automotive manufacture may recognize Boeing's strategy, however. Boeing's approach is Lean—it is based on smaller batch transfers with smaller aircraft. It means nonstop flights to and from smaller, less crowded, traditionally secondary airports, or minor hubs. Rather than overload hub airports, it assumes that the correct exploitation strategy is to avoid the hubs altogether. It might produce higher cost-per-passenger-mile numbers, but you can't help wondering—over the next 40 years, will the ever-wealthier customer be prepared to pay more for a better experience based on nonstop flights? Is it total journey time that ultimately defines the airline industry? How much are you prepared to pay to fly nonstop and save time? Recently I paid almost 50 percent more just to fly direct from Seattle to Washington, D.C., avoiding a connection and a lengthy delay!

My money's on Boeing!

●　●　●　●　●　●　●　●　●　●　●　●　●　●　●　●

Ironically, since announcing the 787 and building the first prototype, Boeing has hiccupped on manufacturing the plane. In 2012, deliveries are running several years behind schedule with only the first few planes delivered to the launch customer, ANA. Test flights are still seen regularly over Seattle and the surrounding northwestern region of Washington State. Meanwhile, the three Arabian Gulf state airlines, Etihad, Emirates, and Qatar have built a business operating megahubs from their home bases in the Middle East, using the A380 aircraft. The outcome seems to be that the market is big enough for both Boeing and Airbus to win, and by staying out of each other's way, they have both made significant progress.

RTFM

February 2000

Whose Time is Really Being Wasted?

I was driving back to Dallas one afternoon last week from a client site in Las Colinas. It was mid-afternoon, and as I swung on to the freeway, I was *buzzed by an alien*. The alien had big hair, a black rock-band T-shirt, a pale complexion, and it was driving a Honda Civic, pre–'97 three-door model. The Civic had been lowered and stiffened, the wheel arches flared to accommodate the wider tires, the paint job had been redone a sparkling, multi-coat metallic, giving it a tinsel glow. The back shelf was decked out with Bose speakers, no doubt blasting a piercing thud that otherwise couldn't be heard above the drone of the Janspeed exhaust. The license-plate holder declared the owner to be an alumnus of a local Texas university. Finally, what really caught my eye were the six-inch-high letters **RTFM** hand-cut from vinyl, proudly displayed top-right in the rear window. A little private joke. Well, it made me smile.

As the Honda disappeared slowly into the distance, I thought, "Huh, Tech Support, poor guy! It probably gets to him after a while." Much of what I knew about software and customer problems and technical support started mulling through in my mind as the car inched along on the LBJ Freeway. I got to wondering, just whose time is being wasted, anyway?

The Support Guy's Point of View

RTFM! The cry of the beleaguered tech support agent who has just had yet another thick-headed customer on the telephone. Will these people never learn? Gee, Users! Who needs them? The only thing worse than Users is No Users at all. Right? Wrong!

The support guy believes that his time gets wasted by the customers who should have read the manual.

RTFM! Is the cry of an industry that remains in denial about what it continues to fail to do properly. That failure isn't a failure in support. No. God love those guys who spend their lives answering the telephones. No. That failure is in Design.

The Customer's Point of View

The Customer had a problem to be solved, or a goal to achieve, or just a pure unbridled interest in new technology. The Customer shelled out his hard-earned cash to your company—not someone else's company—YOUR company. He chose YOUR product. She believed that YOU were her friend, and that in exchange for a fair sum of money, YOU were going to deliver something of use and value.

Imagine the Customer's disappointment when that new software let her down. She just couldn't figure out how to use it. It didn't deliver. So, she lifts the phone and calls for help.

The Customer's point of view is that it's her money, it's her time, and it's her right to get a result as fast as possible. YOU, the software vendor or website-service vendor, are wasting her time with your inferior offering.

So, is it fair that she should have to read the manual, too?

It's a Question of Character

I am forever grateful that the world is made up of different kinds of people. Different colors, different types, creeds, cultures, religions, personalities. It would be sadly dull if there were only two or three designs and we were all cloned from that.

There is a type of person who is studied, careful, attentive, and particular. I married one! My wife thinks my ineptitude with software packages is, frankly, inexcusable. In her former life, as an executive assistant to the boss of a large multinational, she was quite a whiz with Lotus Notes, Excel, Word, and so on, and so on. She is great with the fax machine, the answering machine, the programmable phone with voice mail, the VCR, the CD Player, and the iMac. The list goes on and on. She never starts on anything new without reading the manual first. By typical Technical Support standards, she is the model user. Why can't there be a whole world full of users like her?

Frankly, it's not possible! The people with the disposition to read manuals are not often the same ones who will have the ideas that will push your business forward. The reality is that the majority of users have neither the disposition nor patience to read manuals. There are a number of personality types, and in a well-balanced team, they will all be present. Perhaps only a couple of those personality types are manual readers. The others are not.

What makes someone read the manual?

First we have to consider some characteristics of these products. The domestic ones, such as the phone, fax, VCR, and CD player, are completely compelling. Life around the Anderson household would be impossible without them.

As for the software programs, these were absolutely essential for the proper execution of her work tasks. Formal training from IT was given in each one. Sure enough, a manual had been provided and read when required.

However, outside these formal programs, which were essential for her job and used only at work, my wife has mastered nothing else, despite free access to several HTML editors and other application software, she shows no interest. She expects her husband to maintain the family web pages.

There is a simple explanation. Software is simply too difficult to learn. The effort required is too much to justify when a softer option—nagging the husband—is available.

In my wife's case, there has to be a compelling reason to learn the software and read the manual, otherwise she simply shows no interest.

There are times when I will read a manual—particularly when I have invested a large sum. I recently read the manual for my new car—partly because it has all-wheel-drive and I have never had such a car before. I would never, ever read the manual for a rental car before driving off, though. Luckily, cars are pretty easy to use. The variations are minimal. I would never consider calling Ford technical support to say, "Hey, I just got into the new 2000 Taurus and it's raining, can you tell me how to start and stop the rear-window wiper?"

I would, however, consider calling support on a single-usage piece of software. In fact, recently on several websites I've had to do just such a thing.

Trying to Force RTFM

Good support costs. Big companies have found this out the hard way. Try asking anyone from Aldus (the Pagemaker company, now part of Adobe). In order to save money and keep the headline price of the product down, is it not fair to ask the User to read the manual? No!

I've found that firms are increasingly aware of the spiraling costs of technical support. Recently, several clients have asked me to "ensure that the technical-support phone number is very hard to find. Force the user through the FAQ, then the on-line help, then the email support, and if all else fails, then let them call an operator."

Frankly, this appalls me! The client is willing to risk destroying the customer's respect for them, their website, or their product. They will risk their brand simply because they aren't willing to pay for good-quality design upfront.

Designing out Problems is Better than RTFM

When I designed products at MDi Systems (in the early 1990s), I was often out doing pre-sales, evangelizing the product.

Several potential clients would say to me, "What are you doing about technical support?" My reply was simple. "We design it out." I would expand on this. "My job, out here in the field, is to understand what you need our product to do. If I get that right, then the product will do it for you. You won't need support. You asked me earlier why our product was cheaper than some of our competitors'. Well, frankly, there is an easy answer to this. We spend less on support costs and we pass that savings on to you. With our competitors, you're paying for all that technical support hassle up front as part of the price."

What Price a Good User-Interface Designer?

How much do you need to pay a good User-Interface Designer? Probably twice to four times the number you first think of.

What is the Value of a Good Interface Designer?

Anecdotal evidence from my past employment, at firms like Rombo and MDi Systems, has shown me that improved interface design can cut technical support calls by at least 50 percent. What price a good Interface Designer now?

The Customer's Time

Remember, it's always the customer's time that is being wasted. It's also his blood pressure that is high, and his frustration that is important. Not YOURS! Unfortunately, too few companies force the engineers who build the software to also answer the telephones, but the real baddies in all of this are the managers. It is imperative that any new software, whether product, or website service, be properly designed by an experienced Interface Designer. Relieve your Technical Support people and put the cry of **RTFM** in the bin. Design out Technical Support through improved usability!

Recognizing Tribal Behavior

7

THE HISTORY OF THE AGILE MOVEMENT, AND THE ENTRY OF LEAN AND KANBAN INTO THE SAME WORKPLACES, IS A STORY OF TRIBES. THAT'S NOT SURPRISING, BECAUSE TRIBALISM IS ABOUT PEOPLE—ABOUT RELATIONSHIPS, AFFILIATION, MOTIVATION, LOYALTY, AND LEADERSHIP. INTRODUCING a new methodology in the workplace inevitably leads to the challenge of managing tribes.

Agile encompasses flavors such as Feature-Driven Development (FDD), eXtreme Programming (XP), Crystal Clear, and Scrum. For years, affinity groups associated with such methods squabbled among themselves, while together squaring up to the common enemy from the past—traditional software engineering methods and heavily planned, commitment-based, phase-gate project management and governance frameworks. Recently, some members of these Agile tribes have chosen to go to war against the Lean and Kanban communities. This tribal behavior is entirely predictable. The actions of leaders in the Agile community can be easily explained when viewed through a tribal lens.

In 2004, I described Ray Immelman's *Great Boss, Dead Boss* (Stewart Philip International, 2003) as the most important book I would read all year. It had a profound effect on my work and on my approach, several years later, to building the Kanban community. Understanding tribes and the wider field of sociology is now a vital element for someone who seeks to be a successful change agent, consultant, advisor, coach, or methodologist.

Understanding and respecting people isn't merely about recognizing the field of psychology as critical. It is important that we recognize that being human means that we are herd animals and that tribal behavior is core to our very humanity. Leading, managing, and inspiring knowledge workers cannot be done effectively without an understanding of the tribal nature of humans.

Being "Too Nice" (to the Workers)

Thursday, December 18, 2003

• •

His performance as a manager was superior and a considerable improvement over his predecessor. So why get rid of him?

• •

MY FATHER HAS BEEN STAYING WITH ME FOR THE HOLIDAYS. IT'S HIS FIRST TRIP outside Europe. I've been trying to spend as much time with him as possible; I've had a couple of chances to do that this year.

Back in September, he was reminiscing about his short spell as a factory line manager. He worked for 30 years at the Nobel's Explosive Company. Few people remember that Alfred Nobel[1]—famous for the Peace Prize (and others)—was the inventor of nitroglycerin, and later, dynamite. After scouring Europe looking for a manufacturing site, he received permission to build a factory on seven square miles of sand dunes on the north bank of the river Irvine in Scotland, at Ardeer, where the British Dynamite Company opened in 1871. More than 100 years later, the plant, now part of Imperial Chemical Industries (ICI), was still operating as Nobel's Explosives, mostly producing detonators for the mining industry.

For one year of his career, my father managed a manufacturing line staffed by 300 women employed to hand assemble detonators. After only a year in the job, he was sidelined and moved into a teaching job in the firm's apprentice school. There he stayed for five years, as punishment for being "too nice for the job." During his tenure as the line manager, he built a good rapport with his staff. He averted three strikes and increased the line's productivity while avoiding lost production due to threatened industrial action. In Britain in the 1970s, strikes were common and the unions strong. Managers who were good at industrial relations were few and far between. His performance as a manager was superior—and a considerable improvement over his predecessor. So why get rid of him?

He didn't behave like a manager! He didn't behave with the expected social norms of the management tribe. He was considered too close to his workers; this behavior was alien and threatening to other managers. Rather than reward him for achievement, they ejected him from the tribe to remove the threat.

1. http://www.nobel.se/nobel/alfred-nobel/biographical/timeline/

Sense of Belonging

Saturday, May 15, 2004

• •

I admit that I rallied the team with pride and created
an internal rivalry with other teams and business units.

• •

IT'S NOT UNCOMMON FOR TEAMS WITHIN BIG ORGANIZATIONS TO FEEL LOST.
They sit and wonder about the meaning of their work. They have no idea how it
is aligned with the corporate goals or how they can positively influence the stock
price. Many big companies have lost their way and lack vision or leadership at
the top. I worked for one in which the two top leaders were later dismissed and
investigated by the tax authorities for illegal dealings. Allegedly, they were more
concerned with their own wealth than with performing their fiduciary duties.
In such circumstances, it can be hard to motivate a team of developers. Many of
them need to frame their work within a bigger picture, with a goal or vision in
which they can believe.

At Sprintpcs.com, my boss handed down to me his own localized dream
of a future for the telecom company as a major player in the wireless Internet
market. He called this vision "the Mobile Portal." My team (which included some
innovative thinkers such as Martin Geddes[1]) then turned it into a plan for a
platform for mobile web services—an operating system for the wireless Internet.

But all of that came later. In the beginning, I inherited a team with a portfolio of lackluster projects that were going nowhere, bogged down by requirements
paralysis. I needed to lead and motivate this team
to raise their performance to a level that would enable us to move forward with the grand vision that
was emerging from my strategists. So I tied the
dream to a tangible sense of identity and branded
it with an identity and word mark that I had created for us by a friend on the User Experience team.
The symbol signifies "the knowledge of the world
in the palm of your hand"—though some critics
said it looked like a broken propeller. As a brand

Figure 7.1 Mobile Portal identity mark

1. http://www.telepocalypse.net/

mark, it is not world beating. It doesn't work well in inverse colors; it's too complicated, and it doesn't work so well in black and white, either. Compare it with the brands I've created since, such as Agile Management, the David J. Anderson & Associates logo, or Blue Hole Press, and you'll see that I learned a lot about branding from the experience with Mobile Portal.

To be sure, identity marks within a larger organization are divisive. And I admit that I rallied the team with pride* and created an internal rivalry with other teams and business units. Although this had a local positive effect, later it came back to bite me (politically). We were determined to show the IT division that we could develop software faster, better, and cheaper than they could. I cast the bogeyman (the enemy) as another part of the company. Later that changed, and the enemy was recast as other portal companies—AOL, Microsoft, Yahoo! Meanwhile, we started a trend with our identity mark: Other groups in the company began to develop their own logos. About a year later, the Marketing Communications unit issued a company-wide memo banning internal identity marks as divisive. And in fairness, they were. *Mea culpa*—I had been found out!

Nonetheless, my team rallied around the vision for Mobile Portal and they rallied around our adoption of the Agile method, Feature Driven Development. They built a sense of identity with our team logo. They printed their names on cards with the logo on it and displayed them on their cubes for all to see. They were learning a new professionalism as well as learning that projects could be delivered on time and on budget and with the agreed-upon functionality. We even delivered one project without any known defects. They were building a sense of pride in themselves and in their team. They were proud to work for Mobile Portal. For all of us, it was a crucible experience—a coming of age as engineers. For me personally, it was a huge roller-coaster ride politically, but my team were largely sheltered from those issues.**

It's a tribute to how good it feels to work on a great agile team—and how good it feels to work with colleagues whom you know well and you have come to trust through the experience of delivering for each other—that the team, more than two years after I left, still feel that sense of identity and belonging. As the months pass, they are breaking apart—set asunder by corporate reorganizations and better job offers outside the company. But they still want to be together, and as a result, they started their own Yahoo! group—an exclusive club for just the 30 or so of us to keep in touch as the years pass—and they still identify with their team brand. How cool is that?

• • • • • • • • • • • • • • • • •

* Writing in *What Is This Thing Called the Theory of Constraints and How Should It Be Implemented*, Eli Goldratt suggested that emotional resistance should be trumped by a stronger emotion. In the more recent, *Switch*, by Chip and Dan Heath, the authors argue that people must be motivated by emotion in order to change behavior. Logic does not motivate—only emotion motivates. Many managers resort to fear as the simplest choice of emotion to motivate changes. This is a poor choice that is without merit and it is cursed with many negative side effects. A strong alternative is pride. Pride is a strong, and largely positive, emotion. It is popular among military leaders and sports coaches. As this article shows, it is not without its challenges. Hidden in this article is another strong emotion—hope. I gave the team the hope that the Mobile Portal would lead the business to a new era of wealth and success, and coupled that to the pride they would feel not only to be part of it, but indeed, to be the vanguard of it. Hope is a strong positive emotion, popularly used by politicians to motivate support. Barack Obama invoked it to earn his election to President of the United States, John F. Kennedy to send men to the moon, and Martin Luther King to bring civil liberties and equality to America's black population.

• • • • • • • • • • • • • • • • •

** Years later I am still in close touch with many members of this team. I still receive email from some of them out of blue after years have gone by. The common theme in these emails is that Mobile Portal was the best time of their career and it has taken them many more years, and several more jobs and managers to realize that it is seldom so good. This, I believe, is the sad lament of our industry. We pine for the halcyon days of an 18-month to two-year period when we loved to come to the office and our work was both inspiring and inspired!

Another Tale of Belonging

Monday, May 17, 2004

•••••••••••••••••••••••••••••••••••

In such a structured framework, it was easy to make
rapid, accurate, bug-free progress. It was good to be
alive. It was good to be in Ireland in the bubble.

•••••••••••••••••••••••••••••••••••

ANOTHER TALE OF BELONGING DATES BACK TO THE SUMMER I WAS WORKING IN
Dublin, Ireland. It was 1999—a great time to be alive. There was the economic
bubble. It was a great time to be in Ireland. The '90s had been good to the
country. The combination of a pro-business tax regime, demographics that
were producing a large number of young graduates, inward investment, and the
spoils from years of EU handouts were turning Dublin into a boomtown. It was
impossible to get service in a restaurant because all the would-be waiters were
working as web developers.

I was working at a company still known as Trinity Commerce, though it
was already part of Telecom Eireann, which was due to be sold off by the Irish
government in an IPO. This was where Brían O'Byrne and I first created the
state machine execution engine for the user interface layer of the architecture
that was inspired by Ian Horrocks in his book *Constructing the User Interface
with Statecharts* (Addison-Wesley, 1999). This architectural innovation is truly
at the root of this story.

Earlier in the summer, I had analyzed the scope of the project using Peter
Coad's method incorporated into Feature-Driven Development, and sized it as
153 business logic features. The business logic was being written in PL/SQL
for Oracle 8 by a team of Oracle consultants and contractors. I had no control
over that technical and architectural choice. Naturally, I would have preferred
to develop the business logic in Java and use a simple persistence layer tool to
store data in the database. Nevertheless, we ran the project as an FDD project.
A team of seven people spent 11 weeks building the business logic. They were
very professional. Everything got designed, tested, and reviewed. And the quality
was very high—only five bugs in total. Meanwhile, the customer had refused to
de-scope some of the challenging, non-functional user interface requirements.
Their specification was for a GUI (fat client) app and it would not run in a
browser. We pleaded with them, but they insisted that it must work in 3.x web

browsers. Their requirements were effectively mutually exclusive, but logic did not prevail. They refused to budge. In the end, we were forced to build the state machine engine to track the state of the user interface for each user session. This caused a delay in the development of the user interface layer of the system—a long delay. It took 12 weeks to build the first state machine engine. However, the advantage, if there was one, was that we had a complete user interface design and a full statechart model before we started coding the actual application interface.

We were in a position where the project wasn't yet late. The business logic was complete, but almost no user interface existed except for the infrastructure of the statechart engine. We had nothing to show anyone.

Remember, it was the bubble! You can't get served in a restaurant, and hiring developers is difficult. To fill the skills gap, the principals of the company had visited a local college and hired five fresh graduates. We had three of them on our team. Today, these people are probably all experienced professionals, but in those days they were as green as the grass rustling in the fields near our office in Sandyford Industrial Estate, in the suburbs of Dublin. So we were frighteningly close to being late and half the development team was straight out of college.

In order to make significant progress, we agreed that we would work a weekend. Hey, I was a contractor, I got paid by the hour! I agreed to buy the pizza out of my own pocket as a gesture to the team. There were seven team members in all, including the two guys who coded the framework. In addition, Brian Murray, one of the business logic team, agreed to come in to deal with any issues we might find in their code.

Brian soon got very bored with his weekend. After all, the business logic was bulletproof and the user interface had been fully designed with statecharts. Every single call we needed from the user interface to the business logic already existed and it all worked. Brian had little to do but wander around and gaze out of the window. Luckily, he is gifted in a typically Irish way and kept us all amused with jokes and idle chit-chat.

The application was being deployed in Oracle Application Server, which predated EJB 1.0. It was a whole different product in those days. One peculiarity was that it had only a single global logging queue, and we relied heavily on logging instrumentation for debug information. We had seven people writing View and Controller classes—each one of them with a piece of the statechart model to code. In order to test, a developer had to obtain a "lock" on the app server so that no one else would run their code and corrupt the log file. The team all sat together in a U-shaped desk space in a corner of the office. Each member was about four feet from the next. The "lock" was obtained by asking, "Can I clear the log file?" and then waiting for an acknowledgement from the other six members of the team. By mid-afternoon on Saturday, this call had shortened to a shout: "Clear the logs?" Everyone's head is down at their keyboard, coding, and every few minutes, any conversation among them was interrupted with that cry: "Clear the logs?" Indeed, every few minutes a new feature was being tested. The velocity of the team was tangible and directly measurable by the rate of cries of "Clear the logs?" Team morale was high. Everyone knew progress was being made. They could hear it. They could see it every time they ran a test and checked that log file for the outcome. In such a structured framework, it was easy to make rapid, accurate, bug free progress. It was good to be alive. It was good to be in Ireland in the bubble. And it was good to be on the user interface development team of the EIBS project that sunny, long summer weekend in Dublin.

On the following morning, Brian was wandering around the office building bored beyond belief and feeling decidedly spare. Thanks to the wonderful quality of the code he and his colleagues had written, he had nothing to do, so he felt left out and abandoned. He really had nothing to do and he couldn't help us. He was a PL/SQL programmer. The user interface was written in Java. There was nothing he could do. He took to raising his arms and shouting, "Clear the logs!" "Clear the logs!" After a while, we called him over and made him an honorary member of the user interface team. He was now officially allowed to cry, "Clear the logs!" whenever he liked. He was one of us. He belonged!

● ● ● ● ● ● ● ● ● ● ● ● ● ● ● ● ●

*The EIBS cost in excess of two million Irish pounds—about two million US dollars at the time. It was the biggest project Trinity Commerce had ever undertaken. It was their baptism into the telecom industry. It was completed on time and on budget and it included all the exceptionally challenging non-functional requirements that had added significant technical risk. It was defect free. For many who worked on it, it remains one of their fondest memories of that time and of their early careers. Brian Murray is now a process engineer and coach working in Ireland. He remains a friend and still follows my work closely.

Tribalism

Tuesday, May 18, 2004

● ●

I've observed that with some, the tribalism associated with XP goes further. It has, for some, become a religion—a blind faith, a belief without objectivity.

● ●

A SENSE OF BELONGING CAN GO TOO FAR, AND THEN IT BECOMES PROBLEMATIC. It goes too far when it becomes tribalism rather than merely affinity and shared experience. The human being is a naturally tribal animal. Tribal systems seem to have evolved for our survival. However, deep in our genetic makeup, there seems to be a core conflict. Study of populations of chimpanzees and other great apes have shown that women are genetically programmed to breed-out—to maintain the breadth and health of the gene pool. Men, on the other hand, are genetically programmed to protect their own genetic code; this manifests as a tendency to beat the livin' bejeezus out of anything carrying a significantly different set of genetic code. So women like to make love with other tribes, while men like to make war against them. We are genetically programmed this way. This is a core conflict for which I doubt the Theory of Constraints can find a solution ;-). In modern society, most of us ease our inner tribal anxiety with sport. In the UK, and much of the rest of the world, it is football (soccer); in the United States, it is college and professional sports—anything will do, baseball, basketball, football (*sic*), and so on. This ritual of watching sport is usually accompanied by a spouse rooting for the other team just to annoy us (men).

However, we see tribalism elsewhere in society, too. I get asked about it all the time . . .

"Anderson, eh? What's your clan?"

"Anderson!"

"Hmmm. That so! You have your own clan! Do you have a tartan?"

"Yes!"

"A kilt?"

"Yes, but I never get a chance to wear it in the United States."

Political scientists will tell you it's a 101-first-year college credit to understand why politics tends to polarize into two camps. In the United States, you are either a Donkey or an Elephant or a laughingstock Naderite who "wasted his vote." In software lifecycle processes in this modern decade, you are either an eXtreme Programmer or you are a traditional software engineer.* Or so a large element of the XP camp would have us think. The truth is that this form of tribalism is as useful as 30,000 foul-mouthed neds shouting racial abuse at a black player from the terracing of a Glasgow football stadium.

Asking, "Are you extreme or not?" is not the right question. Asking, "How extreme is it?" is still the wrong question. The right question is "What is the best thing to do to achieve the goals of this business?" The answer will vary according to the business—both the business model and the maturity of the industry. Understanding the principles behind eXtreme Programming is what is fundamentally important; then, knowing when to apply those principles appropriately is what follows.

If you must insist on being tribal, and if some modern form of painting your face with blue, wrapping your body in ten yards of plaid in a set pattern, and playing the bagpipes to intimidate the neighboring bullies is your thing, remember that your tribe will make enemies by exclusion.

We don't fight wars over software process. There is no contest to win or lose. When emotions and a sense of belonging are more important than the work, we abandon objectivity. I've observed that with some, the tribalism associated with XP goes further. It has, for some, become a religion—a blind faith, a belief without objectivity. When I arrived in Seattle, in 2002, I recall taking one of my development staff out to lunch, intending to get to know him better. I asked what he thought of Agile development and he replied, "Well, I'm an eXtreme Programmer, so . . .." Later, I relayed this tale to our CFO, who replied, "So, are you telling me that we need to rehabilitate this person?"

• • • • • • • • • • • • • • • • • •

*Note: Scrum had not grown to prominence and dominance in the Agile community at the time of this writing. My concerns about XP from the early part of last decade could now be amplified ten-fold with regard to tribal developments in the Scrum community.

XP as Vanguard of the Revolution

Thursday, May 20, 2004

••••••••••••••••••••••••••••••••••••

Where XP seems to have scored is with its attractiveness to developers. It appealed to their psyche.
It is almost a side effect that it happens to be more
productive in many circumstances.

••••••••••••••••••••••••••••••••••••

MY BLOG ON TRIBALISM SEEMS TO HAVE CAUGHT THE ATTENTION OF AT LEAST
a few notables from eXtreme Programming community. They expressed their
wish to be known as people from the "XP School of Thought."

It seems that just using XP as an example was enough to rile up some of the
XP tribe; evidently they feel attacked or threatened by the suggestion that they
are a tribe or a religion. Ironically, by doing so, they are demonstrating just that!

So now it is time to give credit where it is due.

eXtreme Programming deserves credit for its leadership—for its role
as vanguard of a revolution. I'm sure that Kent Beck, Martin Fowler, Ward
Cunningham, Ron Jeffries, and others did not set out to create a cult, a tribe,
or a religion. Their intent was to preach a better way of building software and
to show that software development could deliver value and delight customers.

Where XP seems to have scored is with its attractiveness to developers. It
appealed to their psyches. It is almost a side effect that it happens to be more
productive in many circumstances. It was XP that gave the Agile movement
momentum. In the popular press, such as *Business Week*, you never see "agile"
mentioned, you see "extreme programming"* used as synonymous with the
whole Agile movement. The enlightened among us know that is wrong. But in
my view, it is all right. If the wider world wants to call it XP because it gives them
a handle with which to grapple, that is fine. I'm not in love with a brand. Like
many other thought leaders in this space, what I want to see is a more economically effective software industry.

In the longer term, political devices such as polarization of a debate among
artisans or intellectuals will fade. There is no need to demonize an enemy when
the enemy is already you. So, too, will branding of prescriptive solutions fade.
What will prevail are the lessons and the principles. We are only just beginning
to come to grips with these, to extract from the prescriptive solutions what is

truly useful. I believe that this shows a growing maturity. Perhaps we can leave blind faith and tribal affinity behind, and adopt a more scientific approach to achieving the goal of more economically effective software production.

In years, or probably decades, to come, software developers will accept the humanity in knowledge work as a basic tenet, and understand that, to compensate for human frailty, it is necessary to do things in small batches and to focus on high quality and highly effective communication. All of these things will be taken for granted. At the root of this revolution in how software is done, eXtreme Programming will have been first and foremost and above everything else the kernel that was once called "Agile."

• • • • • • • • • • • • • • • • • •

* Eight years later, this article seems so dated. Extreme Programming has largely dropped from our parlance and it would be difficult to identify a community. XP evolved into a number of practices with names like Test-Driven Development, Behavior-Driven Development and variants thereof, as well as the Software Craftsmanship movement. It was tempting to go a global replace on the terms Extreme Programing, XP and replace them with Scrum. In today's world of 2012, Scrum has become the term synonymous with Agile software development. An analysis and history of how that change occurred is truly an exercise for someone else in some other book. It is, however, fascinating to realize how much things can change in eight years.

Great Boss, Dead Boss

Monday, February 14, 2005

∙∙∙∙∙∙∙∙∙∙∙∙∙∙∙∙∙∙∙∙∙∙∙∙∙∙∙∙∙∙∙∙∙∙

This is the book that enabled me to understand social
inertia in business and the management challenge of
leading tribes in the workplace.

∙∙∙∙∙∙∙∙∙∙∙∙∙∙∙∙∙∙∙∙∙∙∙∙∙∙∙∙∙∙∙∙∙∙

GREAT BOSS, DEAD BOSS (STEWART PHILIP INTERNATIONAL, 2003), BY RAY
Immelman, has changed the way I think about organizations and communication
down, across, and up. Management is about many things; organizational
structure and communication are perhaps the two most basic. This book has
made me rethink both of those. It's a book about people—about relationships,
affiliation, motivation, loyalty, and leadership. It's a book that is very applicable
to the Agile community. If you recently went through a merger or acquisition—
or are about to—this story of the merger of two silicon chip manufacturing
companies will resonate deeply with you. This is the book that enabled me to
understand social inertia in business and the management challenge of leading
tribes in the workplace.

Tribalism Revisited

Wednesday, March 23, 2005

•••

If managing knowledge workers is all about understanding people, we need to recognize tribalism as an inevitable behavior. We are genetically wired for tribal behavior.

•••

LAST YEAR, I WROTE A TREATISE ON MY DISCOMFORT AND DISSATISFACTION with tribalism in the Agile community. My feelings were based on the assumption that we are all highly educated and articulate and we should be able to put our tribal past behind us and act objectively for the greater good. Well, I was wrong! So it's time to eat humble pie. Ray Immelman takes a whole different view on tribalism. He believes that it is a force of nature. To deny it is to fundamentally deny the human condition. If managing knowledge workers is all about understanding people, we need to recognize tribalism as inevitable. We are genetically wired for tribal behavior. This genetic wiring is pre-human. It can be observed in our close genetic cousins, such as chimpanzees. It is so old that it's impossible to undo through training and education. Immelman argues that denying our tribal wiring leads to dysfunction. He believes that to improve productivity, we must harness the tribal force in our nature and use it to our benefit.

If you want to understand why eXtreme Programming has been so successful, and why people are so passionate about it, you need to understand Immelman's model for understanding tribes (which I describe in chapter 8.) Meanwhile, we can use Immelman's primary approach to understand what might be wrong with the Agile community as a whole. When the tribes are warring, the way to unify them is not to appeal to their intellect (as I did), but to create a super-tribe to which they all can affiliate—a super-tribe that is stronger than their local tribe. A good example of this is the United Kingdom. I grew up in Scotland, and I identify with its culture and rituals. I feel Scottish, and I would identify myself as a member of the Scottish tribe—an affiliation underscored with a broad brogue accent, a penchant for wearing a plaid skirt to formal dinners, and a deeply wired need to eat haggis around the birthday of Robert Burns on January 25. Most of us Scots also identify with being British and being part of the British super-tribe. It is this super-tribal affiliation that holds the United Kingdom together. (Compare and contrast this with the breakup of some larger Eastern European countries into many small nations in recent years.) In turn, I also affiliate with being European and greatly appreciate the increased level of integration that has taken place in Europe during my lifetime.

If we follow this line of thought, the line of thought that created the concept of Great Britain and, several centuries later, the European Union, how might we go about uniting a fractured Agile community? How do we get the Agile community to speak with one voice, to work together to change software development for the better, forever? We need to create a strong Agile super-tribe. So far the Agile Alliance has failed to achieve that. Its membership stagnated at around 1,000 people. Until the Agile Alliance can deliver on most of Immelman's model of 23 attributes and five tribal dimensions, I fear it is inevitable that it will remain a loose affiliation of rival tribes.

(Hint: As a place to start, try the twenty-second attribute from Ray's model: "A strong tribe has a leader dedicated to the tribe's success—a selfless leader who puts the tribe first before his/her own interests."[1] This would be a good place to start. Can you even name the leader of the Agile Alliance?—I'd have to go look it up.)*

● ● ● ● ● ● ● ● ● ● ● ● ● ● ● ● ●

*Six years later, little progress had been made. In truth, things have gone the other way. While the eXtreme Programming tribe was weakening and fracturing into groups, affiliating with various flavors of software craftsmanship and test-driven development, Scrum came to dominate the Agile movement. The Scrum community represented a strong tribe driven by a strong, and overtly tribal, leader. Reflecting on ten years of the Agile movement, I had this to say . . .

1. *Great Boss, Dead Boss* (Stewart Philip International, 2003)

Reflections on Ten Years of Agile

Sunday, February 13, 2011

• •

The enemy is now within. The enemy is, as Joshua Kerievsky put it, "all the crap I see out there," despite ten years of Agile methods.

• •

I SPENT THIS PAST WEEKEND AT THE SNOWBIRD RESORT IN UTAH, HIGH IN THE mountains above Salt Lake City. I'd been invited by Alistair Cockburn as part of a group of approximately 30 people to discuss the future of Agile and to lay out an agenda for the community in the second decade since the Manifesto for Agile Development was authored in the same venue in 2001.*

Only four of the original 17 authors were present: Alistair Cockburn, James Grenning, Jon Kern, and Jeff Sutherland. However, the community was well represented by Agile Alliance and Agile Conference stalwarts like Rachel Davies and Todd Little; Gordon Pask Award winner, Jeff Patton; Extreme Programming thought leader Joshua Kerievsky; software craftsmanship movement leader Russ Rufer; and tool vendors Robert Holler and Ryan Martens, to name just some of the participants.

The weekend opened with the good news that Ahmed Sidky's family had successfully escaped the turmoil in Egypt. It seemed apt to me that the media was full of stories of revolution as we celebrated ten years since the Agile revolution began in earnest with the creation of the Manifesto.

A very full Saturday was professionally facilitated, and I was amazed that it produced a tangible outcome. A statement will be published soon, which all the attendees stand behind. I certainly fully support it.**

However, the exercise left me feeling that the Agile community doesn't know how to operate when it doesn't have an enemy to defeat or a demon to exorcise. This past decade has seen a lot of positive progress. No new demon has emerged. So the new "10 Years of Agile" statement is unlikely to act as a rallying cry, as the original Manifesto did a decade earlier.

The mission now is incremental improvement. It's evolution, education, and improving levels of maturity, rather than a revolution. The enemy is now within. The enemy is, as Joshua Kerievsky put it, "all the crap I see out there," despite ten years of Agile methods.

The Agile movement lacks an institutional home despite Brian Marick's pleas to create one as he launched the Gordon Pask Award in 2005. The Agile community lacks organized leadership. Sometimes the self-organizing, anarchist ethic just doesn't serve the community well. The mission for the movement seemed to be to defeat the demons of twentieth–century software development processes. With these old ideas now firmly in retreat, there is a need for a new mission. But who will lead it? After a decade, there remains no definition of Agile, no body of knowledge, and no umbrella organization that represents the movement. The Agile Alliance organizes a conference every year and then donates some of the profits to smaller community events around the world. But it has failed to galvanize the movement or give it direction. Community-led efforts to archive the collective community lore, such as the c2.wiki, have largely died away.

So although I feel that my 30 or so peers in Snowbird this past weekend did worthy work, I fear that nothing much will come of it. Who will step up to lead the Agile movement in a time of peace? This decade requires steady economic improvements. The revolution is past. And now the community is left without a strong vision or purpose, without a true goal, without an institutional home, and without a vehicle to provide leadership, nor a leader or leaders to drive it forward.

Yes, Agile Alliance, I am talking about you! Can the Agile Alliance reinvent itself for a new decade, or is it time to say goodbye, farewell, and thanks for all the deployed stories? I don't believe the Agile movement knows how to operate without something to revolt against. Agile came, it served its purpose, it had a positive effect. What next? Perhaps it is time to move on? Perhaps many thought leaders already have?

* http://10yearsagile.org/

** A web search didn't produce a result for a directly posted statement following the meeting. This version by Jeff Sutherland seems to reflect the output of the session but doesn't credit it to the 30 or so people in attendance at Snowbird in February 2011, http://msdn.microsoft.com/en-us/library/hh350860.aspx

Apples, Pears, and
Partial Delivery

Sunday, February 15, 2004

I WAS ACUTELY REMINDED OF WORK THIS EVENING WHILE performing domestic duties—making my daughter's pre-bedtime snack.

The customer was busy munching on a plate of cucumber slices that were left over from dinner. She had placed an order for "apple." As our house is semi-Japanese, we serve Fuji apples, which are rather large. I knew that my customer was incapable of consuming a whole apple, and my agility with a sharp knife is not up to sushi-chef standards, so there was no way that I could peel, core, and slice a whole apple in time to meet the customer's demand for immediate delivery. So I cut out a quarter of it and proceeded to core and slice that, peeling the slices individually (I have no idea if this is optimal, as I have never run a time-and-motion study on apple preparation).

With half of the quarter apple sliced and peeled, suddenly a voice proclaimed, "Finished!" and a translucent orange plate appeared above the sofa, gripped by a small hand. Realizing that this was a prelude to the plate succumbing to the laws of gravity, I stopped what I was doing and went to retrieve it. Then I returned to finish the apple preparation.

Within 30 seconds the small hands were now prying the fridge door open with a chant of "Pear! Pear!" As I finished preparing the apple slices, I offered the plate with a conciliatory, "I'll get you some pear. Why don't you eat these apples?" The plate was taken from me and duly launched across the kitchen floor. My diversion—to try and multi-task and bus the empty plate—had delayed delivery of the apples beyond the customer's threshold of tolerance. The customer had consequently changed her mind. Apple was out! Pear was in!

So here is what I should have done:

When collecting the empty plate, I should have delivered the partial batch of apple. My batch size was small, but not small enough. Partial delivery would have bought me enough time to finish the rest of the batch while the customer was distracted—enjoying yummy apple and not left to think about alternatives. Having delivered the second, smaller batch of apple, I could then have moved on to pears, if, indeed, they were even required.

So, how did this reflect my current reality?

It was only last Friday when I heard a colleague say, in response to yet another impending and impossible deadline, "When the date comes, we will deliver what we have working. We'll build only the most important features first. This will allow the customer to start testing, and buy us time to complete the others. When they log a few bugs, we'll slip all the extra features into the bug-fix build—and, hopefully, no one will be any the wiser."

CHAPTER 8

Managing Tribes

W E CAN IMPROVE THE WAY WE MANAGE PEOPLE BY CONSIDERING TRIBAL DYNAMICS. IT'S IMPORTANT TO UNDERSTAND THAT TRIBAL COMMUNICATION IS EMOTIONAL AT ITS CORE. EMOTIONAL COMMUNICATION TRUMPS LOGICAL COMMUNICATION BECAUSE ONCE EMOTIONAL DISSONANCE occurs, logical points are no longer heard. Any comment that glorifies one tribe in favor of another, or that criticizes a cherished tribal practice, is heard as an attack and meets with an icy reception.

It is not enough just to know that there is a tribe within the organization. To manage in a tribal environment, we must be able to recognize a tribe's members and evaluate the tribe's health. Is it in a downward trend? Is the tribe arrogant and oblivious to reality? Is the tribe strong—stronger than other groups in the company?

Agile methods are rife with practices that seem exotic and ritualistic to non-Agile groups within an organization. We have to recognize the impact of the tribal culture both within the Agile group and outside it. Tribal affiliation strengthens loyalty and openness within the tribe. For individuals, tribal validation is even more important than general validation from a wider audience. What will happen when an Agile development team meets with a non-Agile executive team?

In developing the Kanban community, I've had to carefully balance the positive aspects of tribal affiliation with the potentially negative ones. Developing a Kanban tribe has been a positive component of the adoption and spread of the Kanban Method around the world. However, the Kanban Method is, ironically, designed *not* to create new tribes within an organization. The core principles—that you start with what you do now, and, initially, respect current roles, responsibilities, and job titles—are explicitly designed to avoid creating new tribal tension within a technology business. Kanban seeks to introduce change within the existing tribal structure in a nonthreatening way that allows the tribe's image to evolve over time while it maintains its own feeling of safety.

Five Tribal Dimensions

Thursday, March 24, 2005

● ●

Note that self-worth, or individual value, is measured purely in the tribal context.

● ●

BEFORE WE CAN UNDERSTAND HOW TO HARNESS TRIBAL BEHAVIOR TO MANAGE knowledge workers, we need to understand Immelman's basic framework. The thesis of *Great Boss, Dead Boss* is based on the notion of five tribal dimensions:

Individuals are socially, emotionally and psychologically defined by their tribal membership.

Individuals act to reinforce their security when under threat (Individual Security or IS).

Individuals act to reinforce their self-worth when not under threat (Individual Value or IV).

Tribes act to secure their self-preservation when under threat (Tribal Security or TS).

Tribes act to reinforce their self-worth when their security is not under threat (Tribal Value or TV).

Note that self-worth, or individual value, is measured purely in the tribal context. Individual value increases only if fellow members of the tribe recognize a contribution and ascribe higher value to the person who contributed. For example, praise from a non-tribal member does not count, does not increase self-worth. This is the critical premise of the first dimension—everything is tribal.

Here is a list of some of my own tribal affiliations; they are all entities to which I feel an affiliation, though some are weaker or stronger than others.

- ❑ Anderson Family

- ❑ Anderson Clan

- ❑ Scotland (Scottish)

- ❑ Britain (British)

- ❑ Europe (European)

- ❑ Software Engineering Profession

- ❏ Manager
- ❏ Ex-Microsoftie
- ❏ Feature Driven Development
- ❏ Coad/Color Modelers
- ❏ UML Users
- ❏ Agile Community
- ❏ Ballard Residents
- ❏ Seattle Residents
- ❏ Western Washington Residents
- ❏ Mountain Bikers
- ❏ Skiers
- ❏ Strathclyde University Alumni
- ❏ IEE Members
- ❏ Kansas City Wizards
- ❏ MSF Champions
- ❏ TOC (Theory of Constraints) Community
- ❏ Lean/Kanban Community
- ❏ Real Ale Drinkers
- ❏ Single-malt Drinkers
- ❏ Mixologists
- ❏ Sprintpcs.com Mobile Portal (former leader)
- ❏ and on, and on . . .

Think about it and make your own list. What threatens your security, and how do you react to it? Then ask yourself, how do you increase your self-worth? What makes you feel better or worse? If someone insults you, why does it hurt more or less depending on who it is? How strong are your tribes? What threatens their security? How do you feel about it when their security is threatened? How valuable is your tribe? What would increase or decrease its value?

Samurai or Artisan?

Tuesday, April 12, 2005

• •

The reality of much software development, and a key tenet for some in the Agile community, is that software development is a craft. That puts it clearly in the artisan category. Software craftsmanship would be an artisan pursuit, and its output would be considered works of art.

• •

IN TRADITIONAL JAPANESE SOCIETY, THERE ARE FOUR CLASSES OF PEOPLE: samurai, farmers, artisans, and merchants, in that order of importance or rank. Actually there is a fifth class—the untouchables, who are also unmentionable. This group includes shoemakers and leather tanners, who are not classed as artisans. The Japanese are very particular about anything to do with feet. Feet, after all, walk the dirty ground and the shoes covering them are, therefore, unclean.

(A brief tribal aside: One of Ray Immelman's 22 tribal attributes is that a strong tribe is one in which its members know how they compare to the untouchables. To make the unified Japanese tribe of the Tokugawa Shogunate[1] era strong, it was necessary for its society to have an untouchable class.)

• • • • • • • • • • • • • • • • •

*Reading this seven years later, I suppose we could have predicted that the Scrum community would create an untouchable class. When their leader, Ken Schwaber, added a second "t" to the notion of Scrum-But, and personalized it as Scrum-Butt, he was indicating that practitioners of Scrum-But were, in fact, Scrum-Butts, and therefore untouchable. By marginalizing Scrum-Butts, Ken acted to strengthen the core of the tribe.

I've been wondering how the Japanese view software engineers. Are they samurai or artisans? Technically, any knowledge worker with a university degree is a samurai in modern Japan. Software engineers have a university degree. They are

1. http://www.wsu.edu:8080/%7Edee/TOKJAPAN/SHOGUN.HTM

educated. They are part of a profession. So, they are samurai. We associate samurai as medieval warriors. How did they come to be respectable professionals? By the nineteenth century, Japan was so peaceful and civilized that the samurai were no longer needed as warriors. The pursuit of perfection, core to their Buddhist beliefs, had turned warriors, who were perfecting the art of swordsmanship, into scholars, perfecting arts such as medicine and law. Swords became ceremonial—something that a samurai carries as a badge of office. In today's world, those ceremonial swords—the short and the long one—would both come in handy for hacking up source code and refactoring it.

The reality of much software development, and a key tenet for some in the Agile community, is that software development is a craft. That puts it clearly in the artisan category. Software craftsmanship would be an artisan pursuit, and its output would be considered works of art.

So, if the Von Neumann architecture had been around in the 1840s, before America forced regime change on Japan and opened it up to world trade and modernization, would the Shogun have viewed programmers as members of the samurai? Would they have been rich, well dressed, middle-class land owners with servants, horses, and vacation homes in the country, licensed to carry swords in public, and revered by ordinary members of society as their betters? Or craftsmen, hanging out their shingle**, "Java code—while you wait!"

● ● ● ● ● ● ● ● ● ● ● ● ● ● ● ● ●

** The Japanese word for shingle is, ironically, "kanban" in its original meaning and written in kanji (Chinese characters). 看板

Understand Your Tribe

Tuesday, June 14, 2005

● ●

How secure is your organization? Remember, it is
perception rather than reality that matters.

● ●

HOW DOES TRIBALISM AFFECT ORGANIZATIONS AND PEOPLE'S BEHAVIOR AND
relationships within the organization? Ray Immelman has four simple measures
he uses to create a framework of two 2 × 2 matrices. The first (Figure 8.1) is
based on what he calls security—our personal, individual security, as well as
the security of our tribe(s) (organizations or affiliations). Where does your
organization fit on this matrix? How secure do the individuals feel? How secure
is your organization? Remember, it is perception rather than reality that matters.
If you are only two months from bankruptcy, but no one really knows, or cares,
because they feel the business is "safe as houses," you are in the TS+ category,
even though the reality is TS-.

	Tribal Security Positive	Tribal Security Negative
Individual Security Positive	**IS+TS+** • Process focus • Rules and regulations • Infighting and backstabbing • Lack of innovation or risk taking	**IS+TS-** • Cooperative effort to strengthen tribe • Personal sacrifice • Symbols are more important • Common enemy • Rituals practiced
Individual Security Negative	**IS-TS+** • Resignation from tribe • Antagonism towards remaining members • Tribe ejects individual • Individual acts to harm tribe	**IS-TS-** • Everyone leaves the tribe • Tribe quickly dissolves • Individuals lay claim to tribes valuables • Search for new tribe to join

Figure 8.1   Immelman's security chart

The second set of measures involves what Immelman calls "value." Think of it as self-esteem and tribal-esteem. Where does your organization lie on this matrix? Are individuals arrogant? Are they proud? Are they genuinely better than the competition? If so, your organization is an IV+ organization. What about the tribal-esteem? How strong is your brand? Does everyone in your metropolitan area envy your workforce? Are you a destination employer? If so, you are in the TV+ category.*

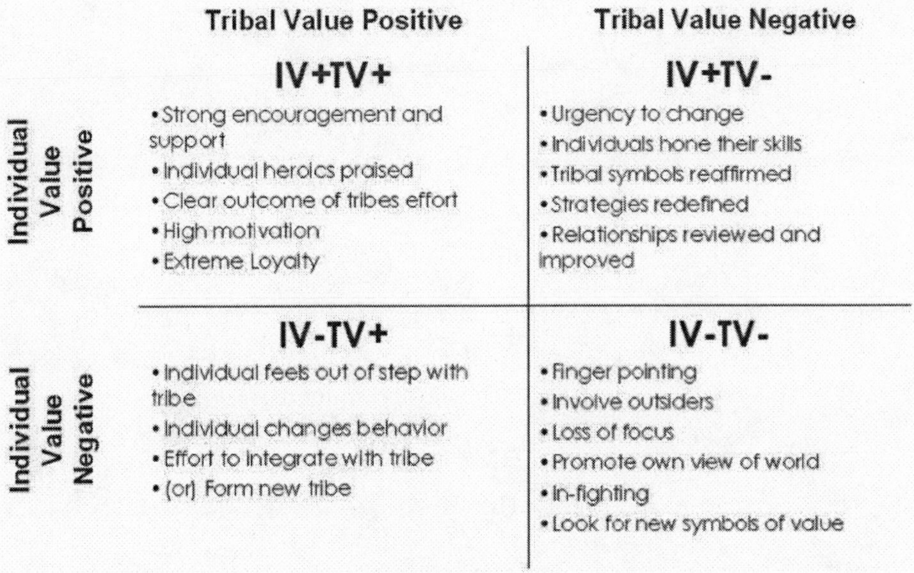

	Tribal Value Positive	**Tribal Value Negative**
Individual Value Positive	**IV+TV+** • Strong encouragement and support • Individual heroics praised • Clear outcome of tribes effort • High motivation • Extreme Loyalty	**IV+TV-** • Urgency to change • Individuals hone their skills • Tribal symbols reaffirmed • Strategies redefined • Relationships reviewed and improved
Individual Value Negative	**IV-TV+** • Individual feels out of step with tribe • Individual changes behavior • Effort to integrate with tribe • (or) Form new tribe	**IV-TV-** • Finger pointing • Involve outsiders • Loss of focus • Promote own view of world • In-fighting • Look for new symbols of value

Figure 8.2 Immelman's value chart

Consider your own organization—your own tribe—against these measures. Now, look at the behavior described in the boxes. Does it feel real to you? Does it reflect back your own brutal reality? Or not? Consider how these matrices would look if you plotted reality against perception. What differences exist? How does this perception-versus-reality gap affect behavior and decision making?

• • • • • • • • • • • • • • • • • •

* Looking back seven years later, we can evaluate behavior in the Scrum community over the intervening years. Ken Schwaber quite ruthlessly removed innovators and dissenters from the Yahoo! group. Some, like me, went quietly, others, such as

Scott Ambler,[1] were much more vocal. More recently, Tobias Mayer[2] made a spectacular departure from the Scrum Alliance and wrote a long blog post showing his antagonism toward remaining members. Immelman's framework would show us that Scrum is solidly in the TS+ category. However, some events over the years provide conflicting evidence regarding individual security. Some behavior, such as the many rejections for certified scrum trainer designation, suggests that those on the inside have IS+. However, Tobias Mayer's story suggests that, at least for him, it was IS-.

1. http://tech.groups.yahoo.com/group/scrumdevelopment/message/17681

2. http://www.infoq.com/news/2010/10/tobias-mayer-part1

Performing a Tribal Analysis

Thursday, June 16, 2005

•••••••••••••••••••••••••••••••••••

Cooperative efforts to strengthen the tribe, personal
sacrifice, focus on a common enemy. That's the
behavior that the company needs to keep succeeding,
to keep winning in the market.

•••••••••••••••••••••••••••••••••••

LET'S TAKE A LOOK AT HOW WE MIGHT USE RAY IMMELMAN'S TRIBAL ANALYSIS
framework to consider a very strong tribe that I got to know well—Microsoft.

Let's first consider Tribal Security. Is the Microsoft tribe secure? Remember,
it's the perception of the tribal members that matters, not external opinion. Well,
it's a big company. Most big companies feel secure. It also has a dominant market
position in its two main markets. It has a lot of money put aside for a rainy day,
and it makes very healthy revenues and turnover. Thousands of people apply to
join the tribe every month. It's a safe bet to say that it has a TS+.

Now, let's consider Tribal Value. Is Microsoft a prestigious tribe? If you
live around Seattle it certainly is. When my wife's friends heard I was being
considered for a position with Microsoft, they all told her how wonderful it
was and how impressed they were. And again, thousands apply to join the tribe
every month—they feel it has value too. If you compare it against its peers in the
Fortune 100, it is highly respected as a business and its leaders are revered. So,
TV+, then.

What about the individuals? We have a great health-care plan, very adequate
remuneration (though it is no secret that some competitors pay higher wages), a
generally strong benefits package, and pension contributions worth real money.
So, IS+. Employees feel pretty secure, which is enhanced by the overall security
of the business. And what about individual value? The company likes to brag
about its great hires. People are referred to as "Microsoft hires." It's generally
considered that the company attracts great people, smart people, competitive
people, ambitious people. A clear case of IV+.

So, why does Bill Gates like to remind the tribe that it is important to "be
paranoid"—that the company could be "innovated out of existence in a 15-year
time period." It seems so unlikely! Well, look at the charts. The IS+TS+ quadrant
is pretty ugly: in-fighting, focus on process rather than customer satisfaction

and delivery, lack of innovation and risk taking, and so on and on. By constantly reminding people to be paranoid, BillG is intuitively moving the tribe to TS-. Look at the behavioral difference in the IS+TS- quadrant—cooperative efforts to strengthen the tribe, personal sacrifice, focus on a common enemy. That's the behavior that the company needs to keep succeeding, to keep winning in the market. This is just one example. It wouldn't be hard to do a full analysis on Microsoft and its business units, but I think such an exercise is best kept internal, don't you?

Clearly, anyone who builds a big business from scratch in his own lifetime is a great tribal leader who intuitively knows how to build a strong tribe. It's obvious that the leadership at Microsoft has done this for 30 years, and it comes naturally to them. Luckily for the rest of us, Ray Immelman has built a framework that we can all use to understand the tribal behavior we see in our organizations, and harness it for our own benefit. Ray teaches us where and how to make adjustments in tribal behavior to drive the desired outcomes and deliver the best returns to the tribe, its members, and its investors.

Decoding Tribal Language

Tuesday, March 28, 2006

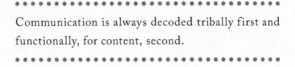

Communication is always decoded tribally first and
functionally, for content, second.

TODAY I MET WITH MY NEW BOSS, RICK MAGUIRE, FOR THE FIRST TIME. I'VE
also been biking in to work again. I don't leave my 1998 model Corratec team-
racing mountain bike lying in the parking lot—instead it leans against the back
wall of my office. So, Rick comes in, and immediately the bike catches his eye,
and he says,

> "So, you are a mountain biker, then?"
> and I reply,
> "No, I'm just a guy with a mountain bike" (albeit an expensive racing model)
> This made him smile. Why did I reply in this fashion?

Quite simply, because his first question was overtly tribal in nature, even
if he didn't mean it to be. Communication is always decoded tribally first and
functionally, for content, second. Hence, he really asked me if I was one of the
mountain-biking tribe. Rick is maybe early fifties, bearded, affable, warm, and
humorous. He's been at Microsoft a long time, and he is entirely comfortable
with his job and his status in the company. He lives on Bainbridge Island, west
of Seattle, and commutes on the ferry to the city every day. About two days per
week he comes in by bike, riding the 14 miles to the office from the ferry terminal. It turns out he does consider himself one of the tribe and was searching for
a fellow tribe member. I picked up on this and my reply communicated that I
am no longer worthy of the tribe. My reply was functional—just the facts—but
equally, it has a tribal decoding of, "No, I'm not one of your tribe."

Why not? Anyone from the original Feature Driven Development project
in Singapore can attest that I was definitely one of the mountain-biking tribe at
that time. Well, no more! The last time I did any serious mountain biking was
on Orcas Island in late summer of 2003. Orcas is famous for some of the best
single-track terrain in the United States. I wasn't even riding my mountain bike,
but rather my hybrid bike (without suspension), complete with a child seat and
a 14-month-old baby in the back. Earlier that summer we'd also done about 12

miles of the Mackenzie River single-track trail in Oregon, too. Since then I've been too busy being a dad. These days my bike is set up like a time-trial bike—with a racing gear block and a big chain ring, slick tires, and configured so that the handle bars are six inches below the height of the saddle. You simply couldn't climb mountains on it, but you can cruise downhill on 15 Ave. NW in the 48/11 ratio at 30 miles per hour. So, no, not one of the mountain-biker tribe anymore; and pretending to be otherwise would be a sham and a breach of trust.

Insulting the Motown Tribe

Friday, February 23, 2007

• •

An appeal to logic will never succeed against
emotional resistance . . . Any response needs to be
emotional in nature and designed to restore tribal
value and security.

• •

MIKE NEISS, WRITING IN THE TOM PETERS BLOG, ALERTED US TODAY TO HOW
the First Gentleman of Michigan and author of *Everyday Leadership*,[1] Dan
Mulhern, managed to insult Detroit's auto-industry workers.

Writing in his blog, Dan comments:

> Because I am married to the great governor of Michigan, I have had the
> chance to be a fly on the wall (generally a quiet and unobtrusive one)
> during meetings with executives from Toyota. These Toyota execs are like
> those I have gotten to know from *Fortune* magazine's "100 Best Compa-
> nies to Work For," in that they get totally fired up when they start talk-
> ing about the culture in their companies. The Toyota folks and the great
> company folks know that "culture beats strategy" every time. They have
> strategies to achieve results. But they know and constantly verbalize that
> the *only way you get results is through people*. For these folks it's not just
> that people are the necessary means, but that people are ends in them-
> selves. It's not just that the employees are there for the company's success,
> but at some really deep level, they believe the company is there for the
> employees' success. So they pay attention to people. They have a *what* of
> results they're pursuing, but they pay primary attention to the *who* and to
> the *how*. (http://www.danmulhern.com/wordpress/2007/02/)

Writing in the Peters blog, Mike comments:

> Daniel Howes, a very good *Detroit News* columnist, took the governor's
> husband to task for "gushing over Toyota's way."[2] He makes the point that

---

1. (Univ. of Mich. Press, 2007)

2. *Detroit News*, 2/27/2007, Automobile section

the governor's husband is sending the wrong message about Detroit, and the auto companies that call it home. Mr. Mulhern wrote that at Toyota the predominant thought is that culture beats strategy every time. As a Michigan resident, and one who has benefited greatly from the auto industry, I question why stating this is being "disloyal." It is the message that Big Detroit Auto has to hear. Loud and often! (http://www.tompeters.com/dispatches/009571.php)

This story is a classic example of how, as Ray Immelman[1] points out, all communication is decoded tribally first, then for logical content afterward. Unfortunately, Dan Mulhern walked right into it. He insulted the tribal value and threatened the individual security of members of the Detroit auto-workers' tribe. Had he been able to communicate the same message in a way that didn't raise the tribal hackles, he'd have made his point, his audience might have taken it to heart, and everyone would have been happy. Instead, his comments have been interpreted as an attack, and there will be no learning.

Mike Neiss continued with this logical argument:

Decades of success have deeply rooted a culture of entitlement at the Big Two. However, a culture built in the '60s doesn't fit the business world today. They are talking the right talk about change at GM, Ford, and the UAW, but they may not be willing to abandon the old to make room for the new. GM and Ford executives continue to ask for the government to address trade policies and CAFE restrictions. The UAW has to understand that benefits gained in the glory days can't be paid for in these difficult times. Employees, both blue- and white-collar, must abandon any notion that they are entitled to lifetime employment and lifetime benefits in retirement.

Organizations that have the passion and discipline necessary to change their culture rely on the truth. And they welcome the truth tellers. In today's *Detroit News*, another fine columnist, Laura Berman,[2] drives home the truth, that we become "enablers" of the current, and ineffective, domestic auto culture. I fear she may have a point.

I grow concerned that a new debate may break out on who is to blame if Ford cannot survive, or GM employment numbers continue to tumble. The question I would pose to our community here is this: Is it possible for large companies with a long history to change their culture?

1. http://www.greatbossdeadboss.com/

2. *Detroit Free Press,* 2/22/2007, Opinion section

Examples? When people offer criticism are they being disloyal? I do hope we can discuss this without turning it into a "who is to blame" thing. Can the truth set the Big Two free?

While everything Mike says here is logically correct, he is appealing to reason when the objection is emotional. An appeal to logic will never succeed against emotional resistance. The mistake was made when Dan Mulhern insulted the Motown tribe. Logical argument cannot repair that damage. Any response needs to be emotional in nature and designed to restore tribal value and security. An appeal to pride, perhaps? A rallying cry that, "We [the Motown tribe] can show those Japanese manufacturers that we can beat them at their own game. Whatever they can do, we can do it better, because we invented this industry, we've been doing it longer and better than anyone else . . .."

Inside-inside, Inside, Outside

Tuesday, August 9, 2005

●●●●●●●●●●●●●●●●●●●●●●●●●●●●●●●●●●●●

Individual value in the tribe is communicated by the
location of your office, not by the color of your badge,
or by the incantations woven into your email alias.

●●●●●●●●●●●●●●●●●●●●●●●●●●●●●●●●●●●●

MICROSOFT HIRES FOR INTELLIGENCE, BUT VALUES EXPERIENCE. ONE WAY OF
measuring individual value in the Microsoft tribe is through the allocation of
office space. Unlike many companies, where the position of an office is associated
with the occupant's position in the hierarchy and amount of power, at Microsoft
it is associated with longevity of service. So, you can have a corner window office
even if you are a programmer, or a user experience designer, but you need to have
about 18 years of service (at the time of writing) behind you.

So, outside offices with a view go to long-service personnel. Offices on the
inside of a corridor where the outside offices have windows and a view tend to go
to newer hires, like me. Meanwhile, offices on inner corridors, where neither side
has any windows—an inside-inside office—tend to be used as meeting rooms,
labs, or are reserved for temporary staff who have to share, perhaps with one or
two others.

So, individual value in this tribe is communicated by the location of your
office, not by the color of your badge,[1] or by the incantations woven into your
email alias. This, in my opinion, is a far better way to express value, respect,
and trust in a healthy way, than the more conventional tactics I describe in the
next chapter.

---

1. http://www.agilemanagement.net/index.php/blog/HR_Myth_4_Tribal_Markings

The Free T-Shirt

Thursday, May 24, 2007

· ·

What my friend fails to appreciate is the tribal
affiliation I feel, and how these shirts symbolize and
communicate that. It's not that they're free.

· ·

A FRIEND OF MINE TEASES ME THAT I'M ALWAYS WEARING FREE CLOTHING TO
work. She suggests that this is how I manage to afford the mortgage on my home
in Seattle—I never buy shirts. I have to wear free clothing instead, like this one
I'm wearing today, the 2007 SEPG conference shirt (on the left).

Or this VersionOne
*(aren't they just the coolest
guys in our community?)*
Agile 2006 rugby shirt
that I'm wearing while
teaching a workshop on
Coad domain modeling at
Corbis this past winter.

What my friend fails
to appreciate is the tribal
affiliation I feel and how
these shirts symbolize and communicate that. It's not that they're free. I'm paid
well enough that I can afford to buy shirts if I wanted to. But I love to wear my
SEPG and Agile conference shirts, and a whole bunch of others, that communicate my passion for and affiliation to the software engineering community.

● ● ● ● ● ● ● ● ● ● ● ● ● ● ● ●

*T-shirts became a core strategy for developing the Kanban community in 2009
and creating a tribal affiliation around the concept of "Yes We Kanban!"

The Value of Tribal Leadership

Originally published as "Antibodies Swarm on the
Agile Business Unit" Sunday, July 4, 2005

• •

Who needs a common enemy when you can have
one down the street in the next building?

• •

In my opinion, the death of Sprintpcs.com was a great loss to Sprint's shareholders—even if many of the employees of the PCS (mobile telephony) business rejoiced at its demise.

There have been reports of managers leading Agile adoptions and transitions being axed. Some of these were reported in the Agile community's literature, others merely communicated by word of mouth. But Agile managers being rejected by a conservative organization unwilling or able to change is not news. I do, however, wonder whether .com (as we referred to our unit) is the first recorded death of an Agile business unit ?

.com was built to be Agile—our vice president, John Yuzdepski, talked about the 50-meter rule and the 5-meter rule. These were the maximum distances that customers could sit from developers, and that developers could sit from team members. Of course, he didn't mean it literally—no one was walking around with a measuring tape. But if you wanted to find an entire co-located business unit, where all functions sat together in a single building, and where on-site customers met developers regularly, .com was it.

.com was unusual within Sprint. It was created with every function it needed to operate. During the Internet bubble, there was a hope that one day it might be spun off as a separate business. As a self-contained unit that didn't rely on other functional groups in the company, it was regularly seen as trampling on others' territory. .com had its own business development group, its own finance team, its own operations team, its own strategy group, its own recruitment, its own facilities management people, and its own IT operations and web hosting team. It relied on the parent company only for money, and for access to a large market of customers. It was created as a stand-alone unit; to give it freedom, it reported to the Chief Marketing Officer, Scott Relf. In turn, Scott reported to the President of the Sprint PCS unit, Charles Levine.

.com had several missions: to build an online retail business for phones and service plans; to provide web-based customer support; and, at Relf's request, to develop a viable strategy for the Internet and so-called "layer 7 applications," such as web- or browser-based applications that run on protocols such as HTTP that use layer 7 of the Internet stack.

The argument that "layer 7 was different" didn't cut ice with the territorial guys from the older telecom world and the network-engineering side of Sprint PCS. They would sit in a presentation and then at the end, stick up a hand, point to the architecture diagram, and say, "So which box will I own?" Yep, if you couldn't mount it in a 19-inch frame and monitor it with SNMP, it wasn't anything a telco guy could understand. The idea that you would build an entire business model on layer 7 of the IP stack, and all you needed from below was an IP connection and quality of service in layers 3 and 4, was bewildering to most of the people we encountered at Sprint at that time.

.com's marketing people were accused of intruding on other marketing groups, such as Business Marketing or Consumer Marketing. This problem was easily fixed—everyone reported up to the Chief Marketing Officer, Scott Relf, and he did what he could to make it clear that .com was to be allowed its freedom. The development teams within .com were seen as intruding on IT territory—except for my Mobile Portal Wireless Application Manager team that was seen as intruding on the network engineering unit known as TASD.

It's easy to look back now, with the benefit of my recent understanding of tribalism in the workplace, and explain what went wrong. Sprint PCS was divided into tribes. The super-tribes of Sprint, and its offspring, Sprint PCS, weren't strong enough to galvanize employees in separate functional business units together. There was an IT tribe, a TASD tribe, a strategic planning tribe, and so on. Everyone was too comfortable. It was the telecom bubble. The stock price had soared to about $70 and then flopped a bit to $45. No one could see the $1.75 low that was coming. No one was threatened. They weren't even afraid of competitors. The whole market was expanding. It was a land grab. Salaries were high. Bonuses were high. Fast-track promotions were the norm. For example, the CTO, Oliver Valente, was in his early thirties and had joined the company only seven years earlier as a mid-ranking engineer. Within Sprint PCS it was IS+ IV+ TS+ TV+, in spades. Recalling the attributes from Immelman's model in the Understand Your Tribe article (page 168), we can anticipate infighting, backstabbing, a lack of innovation, and a lack of risk taking. And that is exactly what happened. Who needs a common enemy when you can have one down the street in the next building? Right?

Into all this came .com—a self-contained, vertically integrated business unit, in a functionally organized company. It was an alien form of bacteria, and the antibodies swarmed on it. .com was its own tribe. It had its own building on Meadow Lake Parkway, in Kansas City, Missouri, away from the Sprint campus in Overland Park, Kansas. It had its own rituals, its own values, its own skill sets and tools—it wasn't a telco! It was a pure-play Internet company. It made software for layer 7. Period.

And it was Agile! It was using Togethersoft's Together product, and Java, and a collection of other IDE and testing tools that weren't the Sprint corporate standard—Sprint was a Rational shop, and still is. It was doing Feature Driven Development at an integrated program management level of competence. It was providing transparency with monthly operations reviews. And, most of all, it was turning out a lot of product.

So what, if there was criticism that a few high-profile projects were late and/or buggy! When taken in the context of the cost and productivity per employee, it outperformed other business units. People envied .com's elitism. They envied its freedom to do its own thing. John Yuzdepski had galvanized a strong tribe. It quickly became the common enemy within. The other tribes swarmed on it and their political maneuvering ensured its death. The end was fairly swift—a decapitation. The VP was removed. His successor was tasked with winding it down and breaking it up—putting back all the integrated functions into their functional silos once again.

Of course, it is all water under the bridge now. Since then Sprint has undergone a major reorganization. The independent PCS tracking stock disappeared and the business was realigned into consumer and corporate units. Now it is going through its own super-tribe problems with a merger with Nextel. I imagine that those who successfully eliminated the .com antibody will find it somewhat harder this time as the Nextel and Sprint tribes fight for power and supremacy in the new, larger entity.

If anything underscores the need for Agile Management at an enterprise level, it has to be the observation that a *Fortune* 250 company can go through three major reorganizations in less than five years, during which time it killed or divested significant pieces of its business and then merged with a major competitor, all while initiating several entrepreneurial ideas and new business units (of which .com was just one). Change is the norm! There really is no such thing as TS+. TS+ is an illusion. Great tribal leaders like Bill Gates intuitively know that they cannot allow the tribe to get too comfortable. Great leadership ensures just enough insecurity to keep innovation and risk taking at the fore while it controls the comfort that leads to backstabbing and infighting. Ultimately, Sprint lacked a great tribal leader, and I believe its stock price continues to reflect that to this day.

More Brains than Agility

March 18, 2005

Stephen Hawking is on record as saying that you can optimize your brain for intellectual capacity or for quick wittedness. Think of it as an athlete—you can have one great, big, muscle-bound monster between your ears, or you can go for a svelte, lithe, quick, agile model with less intellectual horsepower. James McLurkin[1] seems to be one of the former. In fact he's so muscle bound by intellect that even his monthly schedule lacks flexibility.

James McLurkin might be off-the-scale brainy, but he's an old-paradigm thinker. In order to get everything done, he plans his daily schedule to the minute. But his girlfriend complains that he gets up to two hours behind schedule in a typical day. Why? James has no theory of variation in his scheduling. For all his brainpower and focus on complex adaptive systems and robotics, he hasn't discovered the theory of variation, or realized that his daily life suffers both stochastic common-cause variation and uncertain special-cause variation.

He's also got cost-accounting syndrome. He's fascinated with efficiency as a way to give himself more time. James knows that time is his constraint. There are only so many hours in the day, so he doesn't want to waste time doing laundry. The answer—save up six weeks' worth of laundry at a time, and do it all in one night. Yep, he went out and purchased enough clothes to last for six weeks. He also had to get a bigger closet, and presumably, a bigger house for that bigger closet. Yep, James has a lot invested in clothing inventory.

1. http://www.pbs.org/wgbh/nova/sciencenow/3204/03.html

So, he's really smart—right? Well, I wouldn't doubt it. I'd be lucky if MIT let me in for a tour, never mind a Ph.D. But I know that if I need my favorite Helly Hansen (Northwest chic) casual shirt for an evening of beer, sushi, good company, pool, and dancing, I won't have to wait six weeks for the large batch of laundry to process. In order for James's efficient life to work, he must be able to plan with accuracy. There must be low uncertainty in his life. He must not play much at sports. I sure wouldn't let a pair of my sweaty cycling shorts sit around for six weeks before I washed them.

So here is the lesson—large batch sizes may be the algebraic answer to efficiency, but the cost is a large investment in inventory, high carrying costs, and a lack of ability to cope with variety, change, and uncertainty. Not to mention the embarrassment of being unfashionable because you can't afford the replacement cost of such a large wardrobe—after all, to be efficient, you must keep the clothes longer or the cost per wearing will be ridiculous. . . .

On Human Resources

F OR THOSE OF US WHO MANAGE IN FIRMS LARGE ENOUGH TO REQUIRE A HUMAN RESOURCES OR PERSONNEL DEPARTMENT TO MANAGE THE LEGAL OBLIGATIONS OF EMPLOYERS TO THEIR EMPLOYEES, THE POLICIES THAT CONTROL MUCH OF MANAGERIAL DUTIES SEEM COUNTERPRODUCTIVE.

Agile software development has always been about collaborative, cooperative teamwork and organizing individuals into cross-functional teams of generalist workers.

Agile approaches encourage transparency, high levels of craftsmanship, respect among individuals, team work, and a shared responsibility and accountability for success. Yet many of us are forced to work under policies that ask us to assess individual performance and to reward individuals for their contributions.

Over the years, I've posted a series of articles that sought to expose these counterintuitive people-management policies. Exposing the problem isn't offering a solution or a pragmatic means to drive real change. However, several of these vignettes highlight good examples and real, meaningful change at large companies such as Microsoft.

I continue to believe that it is possible to develop a more general theory of people management that is compatible with a holistic, systems-thinking approach, and is better aligned with the value system inherent in Agile and Lean thinking. Respecting people will eventually, I believe, bring about a change in which it's understood that productive, successful knowledge-worker businesses actually manage through policies that respect and motivate the humans creating the knowledge.

HR Myths, #1: Merit-Based Pay

Sunday, November 23, 2003

• •

It is further assumed that hiring managers might not be good at screening candidates. It is further thought that hiring upper-quartile people sets a dangerous "elitist" example . . . In other words, mediocrity is assumed, desired, and managed.

• •

"MERIT-BASED PAY," OR "WE DON'T CARE ABOUT PRODUCTIVITY, LOOK AT OUR Cost Control," is the first in a series of articles in which I seek to blast away the common policies of human resources departments that exist in the software industry and that I believe are doing great damage.

It's typical in the knowledge-worker field for companies to have pay scales and levels; for example, Level 56 = $24,677 to $33,103. Every employee who is at the Level 56 pay grade receives a base salary between the two amounts. In some companies, these scales are lies. Often, the mid-point or median is used to determine what people are actually paid. In this example it's around $28,500, which essentially means that the true pay scale is $24,677 to $28,500.

To determine what the salaries will be, companies' executives decide what kind of performance they'd like to achieve for the stockholders. Next, they decide whether they need upper-quartile, second-quartile, third-quartile or lowest-quartile employees to achieve those goals. They ask themselves, "in order to get the performance we want for our investors, in what range of the pay scale must we pay our employees?" Today, many large companies don't pay more than the third quartile. This essentially says that they want to pay below-average salaries for a given position and experience level and that "average" industry performance is good enough for their shareholders. In this respect, most pay scales are lies, as the upper end of the scale is never actually paid to anyone..

• • • • • • • • • • • • • • • • •

* While I worked at Microsoft (2004-2006) employees who were above the mid-point on the pay scale for their grade were penalized. The policy was that if an employee was good enough to be above the mid-point, they ought to be good

enough for promotion to the next level, where they would once again be below the mid-point. However, some, usually older, more experienced employees, had reached an invisible ceiling. Perhaps a performance review several years earlier had tagged the individual as "not management material." This would have limit that employee to level 64 on the pay scale. Once they were above the mid-point, their bonus-earning potential and pay-increase potential was reduced. Employees above the mid-point, regardless of their contribution over the previous year, got smaller bonuses and smaller pay rises than those less experienced people below the mid-point. This mechanism was used to discourage employees from staying with the company, and instead to seek employment elsewhere. You can draw your own conclusions as to the effectiveness of such a policy.

Hidden here is the classic cost-accounting mistake of focusing on cost. There are numerous examples of differences in programmer productivity being upward of ten- to twenty-fold between a good programmer and a bad one. In fact, some studies say the spread from worst to best can be as much as 50-fold.[1] However, a good manager knows that they need pay only a modest amount more for good people—perhaps the upper quartile for a pay grade. So, why not hire only upper-quartile candidates? Pay 50 percent more and get 1,000 percent better productivity. Hire fewer people and control costs that way. Isn't that a better solution?

My feeling about why this doesn't work in large companies relates to a lack of capability at benchmarking.** It is assumed that hiring managers do not have good metrics or methods of measuring relative performance and hence providing fair pay for everyone of a similar standard. It is assumed that hiring managers might not be good at screening candidates. It is further thought that hiring upper-quartile people sets a dangerous, "elitist" example for engineers, if other jobs in the company are not hired this way. In other words, mediocrity is assumed, desired, and managed. Specific grades are generally awarded based on length of service, experience, or age, as this is assumed to be the most reliable indicator of someone's contribution and their value to the organization.

●　●　●　●　●　●　●　●　●　●　●　●　●　●　●　●　●

** "Benchmarking," or organizational process performance, is a CMMI Maturity Level 4 capability. As there are very few Level 4 or above software-development organizations, it stands to reason that there is an assumption that benchmarking cannot be performed.

1. Reference: Sackman, H. W.J. Erikson, and E.E. Grant, "Exploratory experimental studies comparing online and offline programming performance," *Communications of the ACM*, 1968, 11(1)

So, companies that claim to be meritocracies, in fact, are not. They pay people in a very narrow band. In the example here, you can be sure that everyone in that Level 56 grade is being paid between $27,000 and $28,500, despite the fact that some of them are ten times more productive than others.

One way to break HR Myth #1 might be to refocus the business using a Throughput Accounting focus of productivity first,*** investment second, and cost last. Cost first generates mediocrity and mediocrity results in very poor performance from a software-engineering organization.

● ● ● ● ● ● ● ● ● ● ● ● ● ● ● ● ●

***I now believe that the best way to reward people is by their contribution to managing risk for the business endeavor. A more generalist worker should be rewarded more than a specialist unless that specialist skill is differentiating for the business and is directly generating marginal utility. Greater experience should count, as judgment is an important aspect of risk management. The critical challenge is finding a way to visualize the risks that are being managed, as well as recognizing an individual's contribution to risk management. At the time of writing, I am refining some techniques for this and I hope to publish them in a future book.

HR Myths, #2: Divide and Conquer

Sunday, November 24, 2003

•••••••••••••••••••••••••••••••••••••

HR is thinking about a purely local optima—reduced complaints to HR. They aren't seeing the system as a whole.

•••••••••••••••••••••••••••••••••••••

"Divide and Conquer," or "You Get What You Negotiate!" Many HR managers see it as their job to squeeze new hires on their pay and benefits packages. They see this as reducing costs and directly increasing the bottom line. In some cases, they are incentivized to make good deals with new hires. This can be counterproductive. Once again, it is a cost-accounting driven focus on reducing cost.

In almost all companies, it is specifically against company rules to discuss with your colleagues what you—the knowledge worker—are paid. Why? Simply put, it is a divide-and-conquer strategy implemented by Human Resources. They believe that by enforcing silence with a threat of summary dismissal, they will save the company money and reduce complaints from disgruntled employees. In a few companies, it is also illegal to discuss your pay scale or grade with other employees. This is the ultimate in Big Brother–style control because it theoretically prevents employees from learning that someone doing the same work is in a higher grade than they are. HR believes that this enforced silence reduces complaints, saves the business money, and makes employees happier.

From reading the work of Steve McConnell and Paul Glen, I believe they would both disagree with this. HR is thinking about a purely local optima—reduced complaints to HR. They aren't seeing the system as a whole.

Steve McConnell believes in total transparency in pay scales. The need for transparency has been a common theme in my own writing. Steve walks the walk by providing total transparency in the remuneration process at his company, Construx of Bellevue, Washington.

> *Salary Structure.* We've found that an organization's traditional reward system must be structured to support [personal] professional development goals; otherwise, project goals will supersede [personal] development goals. . . .

Our ladder levels have exactly one salary level at each level. The salary for each level and each employee's ladder level is public information within Construx. Employees have a tangible incentive to reach the next ladder level because they know the salary adjustment that will occur with that promotion.[1]

Note that there is only one precise salary amount per pay grade—no band or scale. There is total transparency in the system, and as Steve describes elsewhere in the book, there is a publicly defined definition of qualifications for each level on their pay ladder. Every employee is able to judge whether other employees are fairly graded because they know them, they work with them, and they will know their skills.

Paul Glen explains in Leading Geeks why Steve's approach is better. In the chapter, "The Essential Geek," Paul explains the elements of what he refers to as the "geek psyche." One of those elements is "a sense of fairness." Geeks love and expect meritocracy.

> Geeks are generally not captivated by money.
>
> Their attitudes to money are much more tied up in their strong sense of fairness and justice. No one wants to be taken advantage of; everyone wants to feel fairly compensated for their value. The passion for reason combines with a strong belief in meritocracy to create an atmosphere where money is a primary measure of the value that one delivers to the organization.[2]

As a manager, I spend a lot of time balancing equity among team members. I make adjustments in pay to recognize contribution and to reflect individual merit. I often get into heated arguments with HR people. Why? Because I'm often trying to clean up the messes they make. Geeks should not be penalized on pay because they are inexperienced negotiators when they engage with HR professionals during the hiring process.

Software development *is* a people business. If you want to exploit people's potential you must keep them happy and well motivated. A transparent, merit-based pay scale is a major factor in the motivation and productivity of knowledge workers.

1. McConnell, Steve, *Professional Software Development*, Addison-Wesley, 2003, page 158

2. Glen, Paul, *Leading Geeks: How to Manage and Lead People Who Deliver Technology*, Jossey-Bass, 2003, pages 40-41

HR Myths, #3: Performance Buckets

Wednesday, March 24, 2004

• •

If you don't want to get that bottom rating, better get in good with the boss, better get noticed, better not give selflessly to the team (or the customer, or the business, or the stockholders).

• •

AROUND THIS TIME OF YEAR, STAFF MEMBERS ARE PAID ANNUAL BONUSES AND they receive merit increases in basic pay. These are based on the results of a performance review that took place in the first quarter. The individual's performance was assessed against some goals, and a rating—usually on a scale of one through five, where five is the lowest—was assigned. For many managers, this process is one of the hardest things they will ever do. Why? Quite simply, the rules in most big companies are not aligned with the best interests of the employees, the customers, or the stockholders.

The first fallacy is the concept of an equal distribution of performance across a team—someone has to get a one and someone has to get a five. The idea is based on the statistical bell curve normal distribution. However, when your statistical sample is (for example) less than 20—as any statistician will tell you—you do not have a basis for a normal distribution.

The second fallacy is that all teams perform equally and that someone from every team deserves a one; and, equally, someone from every team deserves five.

The double combination of these "buckets" for performance rating means that ultimately a manager has to have this chat (or one like it) with some poor, unfortunate employee:

Well, Joe, as you know, we both reviewed your performance and agreed that you have given everything asked of you—and more—to our team's performance this past year. As you know, our team delivered on its promises. In fact, our Agile techniques helped us become the highest-producing team in the company, both in productivity and efficiency and in costs. Your contribution was key to this success. However, as you know, this is a strong team, and all of your colleagues contributed strongly too. You know how the performance rating system works, and unfortunately, after discussions with oth-

er managers in the business unit, I, regrettably, have to give you a four rating. I managed to argue that no one on our team deserved a five. Consequently, and I am ashamed to have to report this, you will not be receiving a pay raise this year and your bonus check is a little light. I want you to understand that this does not reflect how I feel about your performance, and I will try to make this up to you in other ways. However, the company rules force me to make decisions—some that I am not comfortable with. In a team of fast runners, you are being rated as the slowest this year. I hope next year it will be different.

What is the effect of this? In the worst case, Joe goes out and polishes up his résumé and he leaves the company soon afterward. In the best case, he becomes defensive. He doesn't share so readily with the team. He is keen to be given individual credit for his contribution. After a few years, the whole team is acting the same way, and they are no longer a team, but a group of fearful and defensive individuals.

Existing HR policies discourage teamwork, discourage knowledge sharing (death in a knowledge work industry), and encourage sub-optimal—even divisive—political behavior. If you don't want to get that bottom rating, better get in good with the boss, better get noticed, better not give selflessly to the team (or the customer, or the business, or the stockholders). Better still, why not make it difficult for your colleagues to run as fast as you? Why not spend your energy trying to hinder the other runners in the field?

In sports, when a team wins a championship, everyone on the team gets a medal (or a ring) and the bonus payment. It's time we started remembering that when we reward our knowledge workers.

• • • • • • • • • • • • • • • • •

Shortly after this article was posted, Jeff De Luca posted similar thoughts on his featuredrivendevelopment.com site. Actually, the method he describes is not identical to the methods I have seen in two large American companies, but it is similar enough for you to get the picture. He demonstrates a similar concern to mine . . .

H.R. applying the bell curves and normalizing within levels is not just sub-optimal it is often plain unfair. It assumes a related normalization of staff numbers and skills and performances by level across the organization. You can't just say we'll promote 3 associate programmers this year and 2 programmers. What if the associates mostly performed poorly (can easily happen if the hiring was not good). 3 associates get promoted no matter what. What if there is a much higher percentage of outstanding programmers than there is of outstanding associate programmers this year? (again, can easily happen). They are penalized as only 2 will get promoted no matter what.[1]

1. Reference: http://www.featuredrivendevelopment.com/node/691

HR Myths, #4: Tribal Markings

Monday, August 8, 2005

• •

Does anyone really think that full-time employees are any more loyal to their employers than to their own careers and their personal development? In the twenty-first century?

• •

WHY DO HR DEPARTMENTS INSIST ON ISSUING DIFFERENT COLORED BADGES TO contingent contract labor and vendors on long-term contracts? It's clearly tribal. It clearly marks the individual as somehow less worthy. Why? The assumption is that temporary staff are less trustworthy. Hmmm. This feels like it belongs in medieval Japan's early Edo period when Samurai without a master—ronin—were treated with suspicion.

The ronin's loyalty was to himself, as he had no warlord. So, too, is the geek-for-hire contract laborer—loyal only to him- or herself and his or her career. They ply their trade wherever it makes the most economic sense. But does anyone really think that full-time employees are any more loyal to their employers than to their own careers and their personal development? In the twenty-first century? Really? In medieval Japan, loyalty to a group and its master were built into the religion as a core Confucian value. We still see that today in modern Japan; employees are still very loyal to their employers, and in return, they expect a lifetime of employment. But in America? Really?

When I worked at IBM's PC company in the mid-1990s, contractors like me had a yellow stripe on our badges. We also suffered the indignity of a temporary email address. Mine, if I recall correctly, was CONTN664@uk.ibm.com. Furthermore, as I was an untrustworthy ronin, under IBM's strict phase-gate software development lifecycle (a relic of the Fred Brooks days—evidently no one at IBM had bothered to read Fred's retraction), I was restricted to performing only coding and testing tasks; analysis and design were reserved only for full-time employees. Hence, I had to code designs given to me by a full-time employee—an IBMer.

My current employer, Microsoft, too, makes temporary staff wear a variety of different colored badges, segregating them in a strict caste system. And they get special aliases, too. It just doesn't make sense to me. I can see no reason why a temporary knowledge worker should be treated as untrustworthy. They have loyalty to their careers, and their development relies on their carrying good references and respect from one job to another. The temporary staffer has just as much incentive, or perhaps more, to do a good job as a full-time employee.

So, might there be other reasons why HR departments the world over persist in this behavior? After all, it's not just IBM and Microsoft that are doing this. Temporary staff are often paid more than employees, but they generally do not receive benefits such as health care. Sometimes there is resentment about the higher pay rate. At IBM, contractors were expected to pick up the bill if the team went out drinking together in the evening. This wasn't too much of a burden if the group had several contractors who could share the cost. I've also seen managers become irrational when dealing with temporary staff who are doing a good job. They cannot get away from the fact that the temporary worker makes more money than they do. On several occasions I have seen managers eliminate a high-performing temporary worker (by not renewing their contract), only to suffer a delay in a schedule or an outright project failure after the talent has gone. All this to make themselves feel better. As a line manager, I had several temporary staff making ridiculous money during the boom years of telecom. But I simply had to put it out of my mind. They were doing a job that I couldn't easily backfill, and their pay rate was determined by the market. Meanwhile, I got my projects delivered on time.

So perhaps the segregation of colored badges is all about making people feel better or superior? Humiliate the ronin! If this is true, am I the only person who looks at history and finds this idea discomfiting?

When I left IBM to go to Singapore, in 1997, I offered, despite my temporary status, and despite the fact that I had fully met my contractual obligations, to find them a replacement to fill my position. I recommended a college buddy, David Morton. He still holds that job today—many years later—and he is still temporary. About two years ago, I noticed that his email address had changed to david.morton@uk.ibm.com. I assumed that he had gone full time. I learned later that he hadn't. Instead, IBM had finally ended the humiliating practice of giving temporary staff ridiculous email aliases. Kudos to them; a step in the right direction! Software development is a team sport, and productivity is directly related to team morale and the ability of a team to work together effectively. We know that Agile software development rides on trust among team members. Eliminating division and artificial barriers to trust such as different colored badges and funky email addresses is surely an enabler of highly productive teams.

HR Myths, #5: More Thoughts on Pay-for-Performance

Monday, March 12, 2007

● ●

Could it also be true that policies such as pay-for-performance remain ubiquitous because they are deeply ingrained as part of the tribal lore and value mechanism in the human resources tribe?

● ●

THREE YEARS AGO I WROTE ABOUT MY FEELINGS REGARDING PERFORMANCE-related pay, promotions, and bonuses. Although I might deal with the challenge of delivering performance review feedback differently today, having gained more experience at these things, my general conclusion remains the same—pay-for-performance can lead to dysfunctional behavior. Pay-for-performance encourages individuals to look out for themselves, and it discourages teamwork. This is an inhibitor to achieving high levels of team productivity.

At my current employer, our annual employee survey, interestingly, asks the question, "Do we have a good performance-review system?" I wonder if your company asks a similar question of its employees. I can't share our results with you, but draw your own conclusion. Are pay-for-performance schemes universally disliked by a significant portion of knowledge workers?

A couple of years ago, I had lunch with Lisa Haneberg, who has become well known for her books and blog about middle management. Lisa knows a lot more about HR than I do, and she has actually spent time hanging out with HR types at their industry conferences. I asked her whether HR policies come from the tooth fairy, or if she could enlighten me on the source of the ideas. Though she knew more than I, she couldn't point me to original work, or authors, to understand the source.

I expressed my concern that typical performance-pay rules were like cost accounting—somehow left over from a previous era of management. Modern Cost and Management Accounting[1] was created by Donaldson Brown of General Motors to provide a means of making local decisions in mass-production manu-

---

1. http://en.wikipedia.org/wiki/Cost_accounting

facturing. Mass production was based on ideas from time-and-motion studies, described in Scientific Management[1] by Frederick W. Taylor. Cost and Management Accounting gave us the efficiency metric that seeks to optimize utilization at individual stations on a manufacturing line. It's based on the assumption that optimizing locally (for efficiency) will lead to optimal performance for the system as a whole. The concept has largely been debunked by the Lean and Constraints Management movements. Modern, successful manufacturing companies have abandoned Scientific Management, but its accounting method remains the industry standard, and it is widely taught in colleges all over the world. My curiosity revolved around whether HR policies can be traced to a similar pattern of development. Were they created out of the mass-production era paradigm that local optima leads to globally optimal performance? If so, why do they persist in the post–mass production era?

Thomas Kuhn[2] described the inertia that exists within a profession, and why innovation and new ideas often come from outside that profession. For example, Eli Goldratt, a physicist, promoted Throughput Accounting[3] as an alternative to Cost and Management Accounting. Ray Immelman[4] teaches us to look for a tribal explanation first. Immelman's model is simpler to understand but amounts to the same thing as Kuhn's. He suggests that when a tribe has determined its mechanism for establishing the tribal pecking order—the individual value within the tribe—there will be huge inertia to changing that mechanism because it would undermine an individual's ranking in the tribe and cause a loss of individual security. Fearful of a reduction in security and a loss of self-esteem, no one wants things to change. The result is that an established profession has no incentive to change. In fact, it has substantial incentive to maintain the status quo. Hence, Immelman's model predicts precisely what Kuhn observed.

This leads to my observation: If Cost Accounting survives because it suits the accounting tribe, might it also be true that policies such as pay-for-performance remain ubiquitous because they are deeply ingrained as part of the tribal lore and value mechanism in the human resources tribe?

Interesting, then, that when Steve Ballmer was looking for a new manager to run human resources at Microsoft, he chose Lisa Brummel, someone from the marketing side of the business with no background in HR. Did he intuitively know that he needed someone from outside the tribe to shake up HR? Or is

1. http://en.wikipedia.org/wiki/Scientific_management

2. http://en.wikipedia.org/wiki/Thomas_Kuhn

3. http://en.wikipedia.org/wiki/Throughput_accounting

4. http://www.greatbossdeadboss.com/

Ballmer a Kuhnian thinker who understood the need for a non-HR professional to take the reins? Regardless of which, Brummel wasted no time in abandoning the forced ranking, and she severely modified and simplified the pay-for-performance scheme at Microsoft.

Robert Scoble had this to say . . .

> Wow. I missed a HUGE HR townhall, er, employee meeting today (they announced new compensation and review changes). I just got the email from Lisa Brummel and, wow, wow, wow.
>
> Is Lisa reading Mini? Damn straight she is.
>
> This is the "Mini-smackdown" I wanted to see. Hopefully these changes will get us on a more customer-centric path.
>
> One big thing that's gone? Stack ranking. No longer am I judged against Charles and Adam and Tina and Jeff. Now, either I'm doing a good job for Microsoft or I'm not and my review will now reflect that.
>
> I LOVE these changes!
>
> Also, I love the transparency that the Office team is experimenting with (you can see the Office team's ranking, and guess pretty closely what salary each employee there is making).
>
> One thing I love about Microsoft is that we are willing to play with the business and make improvements. For a big business these kinds of changes aren't made easily, nor often, and I appreciate when they happen and the amount of work that goes into making them happen (I know someone in IT for HR, for instance, and he told me about all the work that's going on behind the scenes to change the review system).
>
> Oh, and thanks Mini! These changes are due in no small part to you. Even if you don't get official props in the press releases.
>
> Can one person change a huge company? Mini did. And we don't even know his name.
>
> But, don't miss the work that Steve Ballmer, the leadership team, and Lisa Brummel did here either. Wonderful. Cheers. Now, let's get back to work figuring out how to make our customers lives better.
>
> Lisa announced MyMicrosoft, a series of initiatives that'll make Microsoft a much better place to work.
>
> There's a lot more to what she announced than I'm talking about here, but as I read over the list I'm just astounded.
>
> These are not small little tweaks. They are wholesale changes to how Microsoft treats its employees.[1]

1. Reference: http://scobleizer.com/2006/05/18/missed-big-hr-meeting/

HR Myths, #6: Where Comparative Pay Scales Come Unstuck

Wednesday, March 14, 2007

• •

He believed that you had to optimize each part in the system to be the best. So, if you wanted to concede the fewest goals in the league, you needed to have the best goalkeeper.

• •

HOW DO PAY SCALES COME ABOUT? HOW DO JOB DESCRIPTIONS GET ASSIGNED to a pay grade? How do compensation bands get assigned to those pay grades? And how do HR departments decide how much is fair pay to offer a candidate applying for a job? To be mildly unkind, they are thick as thieves with all the other HR folks! They meet usually twice per year to compare pay scales and compensation bands with other companies in the same metropolitan area or industry, and they use salary surveys and other analyst research data to build a map of what compensation levels are appropriate for particular positions. They accumulate mean and median information, spreads of variation, and distribution curves for different jobs. This data allows them to assess upper quartile, median, and lower quartile pay, often accurate down to the percentile.

At a higher level, usually the executive committee of a business decides which percentile or quartile they want to target for their employees. If for example, an executive committee decides that their strategic positioning is as the cost leader in a market, they might decide that as part of that policy they want to pay only lower quartile salaries. Hence, low cost also means cheap or poor quality staff, presumably leading to poor quality service and products—but at a cheap price.

This would imply that an organization that wants to be merely mediocre would instruct its HR department to enforce a policy of, say, median (or fiftieth percentile) pay for its employees. Congruent, is it not?

STOP RIGHT THERE!

When did you ever meet a strategic planning department that would openly state, "Our goal is to be mediocre!" How motivational do you think that would be? Can you imagine the posters around the office—Master Mediocrity—Strive for Median Performance—Competing is Better than Winning!

And you get my point!

Companies that state, "We want to be Number One in the [insert subject here] market," "We will dominate the market for [insert product] in [insert geographical region]," and so on, are declaring that they want to be the best. So how do you reconcile that with, for example, a recruiting policy of targeting the sixtieth percentile in an industry?

I like to call the proper alignment of goals with recruitment policy the "Kenny Dalglish School of Management." Dalglish was the most successful football player ever to come out of Scotland. Arguably he has to cede the position as best-ever manager to Alex Ferguson, but Dalglish's managerial record is very impressive, having won the English league trophy on several occasions with two different clubs. He took the second team, Blackburn Rovers, from the Second Division to the championship in under five years and thus proved that he hadn't been lucky the first time around with Liverpool. Dalglish had a simple approach to management. He believed that you had to optimize each part in the system to be the best. So, if you wanted to concede the fewest goals in the league, you needed to have the best goalkeeper. If you wanted to score the most goals, you needed the best striker. And so on. He consistently broke transfer records to sign the best players to his team, thereby denying those players to the competition. His actions were aligned with the goals of his employer. Consider this snippet from his Wikipedia page:

> 1994–95 saw Dalglish again break the transfer record, paying Norwich City £5 million for Chris Sutton who along with Shearer formed a formidable striking partnership. He had now spent £27¾ million putting together a squad that could make a serious challenge for the ultimate prize, the Premier League Championship. The challenge came and by the last game of the season both Blackburn and Man United were pushing for the title, Blackburn had to go to Dalglish's former home Anfield with United having to go to East London to face West Ham United at Upton Park. Dalglish smiled as Rovers went 2–1 down to a late Redknapp winner and the news that United had failed to get the result they needed filtered through to him via the radios in the crowd.
>
> At Blackburn Rovers, club owner Jack Walker wanted to win the championship. He hired Dalglish to make it happen. Walker wanted to be number one. Dalglish hired the best by paying the most and delivered the prize.

Now ask yourself this, "Are your employer's strategic positioning and stated goals aligned with the recruitment and compensation policies?"

Further, consider whether one blanket policy—a one-size-fits-all approach—to define the targeted pay band is appropriate? For example, if your employer has a goal to be number one in service, but a goal of lower costs in widget manufacture because the widget market is coming under price pressure and has been commoditized by competitors, would you have a single policy on recruitment and compensation, or would you tailor those policies according to your strategic plan? Mightn't it be better to have an upper-quartile policy for people who can deliver on the goal of being "number one in service," while other areas of the business might rightly have a third quartile policy?

● ● ● ● ● ● ● ● ● ● ● ● ● ● ● ● ● ●

*As a footnote to this article, at the time of writing, Kenny Dalglish is once again the manager of Liverpool Football Club in England. Times have changed. There is much more money in professional football than there used to be. Several English clubs have very wealthy owners with very deep pockets. Dalglish's second tenure at Liverpool has started well. They have already won the trophy traditionally called the League Cup and at the time of writing have reached the final of the FA Cup. They are having their best season in a decade. It seems he inspires players to give their best. This time he has to find a new strategy for success, as he simply can't spend his way to victory. It remains to be seen whether he can rekindle the memories of decades past, and produce a team that can deliver championship-winning performance.

Construction Time Again

Thursday, February 16, 2006

Construction is nothing like software engineering, and construction management is nothing like software project management. Jim Shore has told you so.[1]

I was sitting in a seminar presented by a popular agile tools vendor recently and I suddenly went into an involuntary cringing spasm when the presenter said, "Now imagine we were going to build an agile house." He went on, "If we were building it in iterations, we might decide to build the bathroom in the first iteration, and then the kitchen in the next iteration."

Now, the Agile landscape is changing, but shame on us for continuing with this misleading idea. There might be ways you could bring Agile ideas into house construction, but it certainly wouldn't be by building one room at a time. The answer is to pipeline both the work and the delivery of materials. There is already the Lean Construction Institute[2] and a movement for lean project management in construction led by Greg Howell and Hal Macomber. Truly self-organizing construction can lead to a bazaar or a shanty-town. An in-depth analysis of how unplanned, ordinary development evolves over time is described in *The Structure of the Ordinary* by N. J. Habraken (MIT Press, 2000), while adaptive construction that reacts to changing requirements over time is well documented in *How Buildings Learn* (Penguin, 1995) by Stewart Brand.

From the Agile Manifesto, the only value that immediately jumps to mind as appropriate for home construction

1. http://www.jamesshore.com/Blog/That-Damned-Construction-Analogy.html
2. http://www.leanconstruction.org/

is *Customer collaboration over following a plan.* I fail to see what *Individuals and interactions over processes and tools,* or *Working software over comprehensive documentation,* have to do with construction (as we find it today—and I don't know enough about it to solve those problems for them). I think a vague case can be made for *Responding to change over following a plan,* but who would seriously contract with a construction company without an agreed-upon plan. A garden shed, maybe, but a full house? C'mon, let's get real!

So at this year's Agile Conference's closing dinner, let us make it more than a party and invite Jim Shore to give us two minutes, warning us of the dangers of the construction analogy and then have the audience take a pledge, that although we love, in itself, the idea that agile construction is possible, we recognize that it is a different medium from software and must adopt its own practices that reflect the medium and nature of the work, the tools, the raw materials, and the supply chain.

Everything counts in this difficult job of articulating the evolving Agile story. We just simply need to shake the disease of using construction as an analogy for software development.

[Update: Glen Alleman continues to argue in his Herding Cats blog that I'm confused.[1] Glen adds something to the conversation, observing that an Agile approach to construction ought to be possible and desirable. However, he misses the point that construction and software development are two different mediums. His focus is only on project management, while mine is on the technique of how software is built. You can't refactor a house very easily, nor do you check it in every night before you go home, or test its integration continuously. You specify, architect, design, build, and test software completely differently from those same stages

1. http://herdingcats.typepad.com/my_weblog/2006/02/software_build_.html

of construction projects. Both the engineering approach and, necessarily, the project-management approach are different. Glen asks if I've ever worked on a construction site, to which I'd reply—if he wants us to take him seriously as a commentator on building software—might he like to provide some evidence that he has written commercial software and is knowledgeable in the topic?

Getting back to the agile house analogy, it might be better if the first iteration provided shelter, the second, heat, the third, the non-functional requirement of robustness in all weather, the fourth, water, the fifth, light, and so on. At each stage, the house provides a valued function at the end of the iteration. Staying with the example from the seminar, a bathroom on its own does not function as a house. At best it might be a privy in the wilderness—that bizarre invention known as a backwoods latrine—but it isn't a house.]

● ● ● ● ● ●

So, how many of you aging '80s college/alternative music fans found a hidden layer in this post? There are at least six titles of Depeche Mode songs embedded in the text, as well as the title of the article, which is borrowed from a Depeche Mode album of the same name. This article represents an experimental phase in my creative writing. Looking back, it seems remarkably innovative, and it's hard to imagine myself producing such a piece now.

CHAPTER

10

Understanding Agile

Throughout the last decade, I've worried about successfully scaling adoption of Agile ideas and improving the agility of larger organizations. I always felt that to do this we needed to provide guidance based on a set of principles, and that an underlying model for Agile software development was required to do this. Without it we simply had people superstitiously copying practices without a reference to context and without a means to predict success or judge appropriateness.

Recently, complexity science expert, Dave Snowden and his Cynefin model for understanding the nature of a domain, have become popular among Lean and Kanban advocates. Snowden argues that to scale any technique you must have an underlying model. The Kanban community believes in the use of models and applying the scientific method to process improvement .I have called this a "guided evolution" approach, where each "mutation" of a process isn't random, as with biological evolution, but guided, based on applying a model and predicting a likely outcome for a given change in a process definition.

I advocated for the development of an underlying model for Agile software development at the 2008 Agile Conference in Toronto. Later, I was told that the board of the Agile Alliance had discussed the topic and considered it important, but that no one on the board had any energy to lead a program to develop it. Four years later, it would be unfair to say we are still waiting; more accurately, we could summarize by saying that any understanding that such a model is necessary has dropped out of sight.

Is it any wonder, then, that scaling Agile adoption continues to be a challenge?

Good, Fast, Cheap—Pick Three!

December 9, 2003

•••••••••••••••••••••••••••••••••

The Agile software community is essentially declaring, with its Manifesto, a belief in "The Genius of the AND." Agile software development is all about having it all.

•••••••••••••••••••••••••••••••••

IN *Built to Last* (HARPER BUSINESS, 2004) JIM COLLINS AND JERRY PORRAS describe their conclusions from painstaking research studying large companies that have survived longer than 50 years. They extract the common themes that these businesses have used to be successful. Although there are many useful themes in this book, I'd like to focus initially on "The Tyranny of the OR."

In software engineering, we have been brought up with the notion that there are three main constraints—quality (*good*); time to market, or schedule (*fast*); and cost (*cheap*). It has been widely written that a business must choose only two of these. This has resulted in the flippant saying, "Good, Fast, Cheap—Pick Two." The point is that we must sacrifice one in order to gain the chosen two; for example, good quality, low cost (small number of people and other resources), with a long, slow schedule; or low cost, short schedule, but quality is inevitably sacrificed. This assumption is a manifestation of what Collins and Porras have termed "The Tyranny of the OR." That is to say, you can have good and fast, OR good and cheap, OR fast and cheap, but critically, *not all three*.

Collins and Porras point out that companies that are built to last do not accept the "The Tyranny of the OR," but instead embrace "The Genius of the AND." These businesses simply refuse to accept that it is not possible to do it all.

The Agile software community is essentially declaring, with its Manifesto, a belief in "The Genius of the AND." Agile software development is about having it all—good quality through rigorous testing, reviewing, and learning—fast pace through face-to-face communication, less bureaucracy, and greater acceptance of the risks from relying on tacit knowledge—low cost through small teams of highly talented, empowered, self-organizing, well-motivated generalist developers.

Ensuring good quality does not increase cost or prolong schedules. In fact, it is often the reverse. The reduced failure demand achieved by preventing defects from escaping results in more capacity available to apply to valuable new features and functions. Teams achieve a greater throughput of valuable work, which reduces demand for additional staff and shortens lead times. Among Agile software developers, a core focus on quality and workmanship enables them to break "the Tyranny of the OR."

I believe that 50 years from now, the companies that were built to last in this decade will have demonstrated that they refused to be intimidated by "The Tyranny of the OR," and rather, believed in "The Genius of the AND." By doing so, they embraced Agile software development as a key strategy to having it all.

Trust is the Essence of Agile

Sunday, June 26, 2005

••••••••••••••••••••••••••••••••••••

Audits and arbitration are aspects of a low-trust society.

••••••••••••••••••••••••••••••••••••

BACK IN JUNE 2005, I WAS FINISHING UP MY PRESENTATION FOR THAT SUMMER'S Agile conference. I asked members of the Agile Management Yahoo! group and colleagues working at Microsoft to review my presentation, and in doing so, discovered there are two schools of thought on the topic, *What Agile is all about*. Neither of them matches with my world view.

The first school thinks that Agile is all about feedback loops. It's all small iterations in different shells, from 15 minutes continuous integration to monthly sprint retrospectives. The second school of thought believes that Agile is all about people, and treating people as human, rather than as fungible, interchangeable engineering workstations. Both of these are important aspects of Agile, but for me, they do not identify Agile as unique.

Iterative and incremental development lifecycle ideas have been around since the 1970s. Agile may have added some new tooling twists in recent years, such as continuous integration, but I don't believe this is enough to cite as a unique differentiator. First of all, continuous integration isn't a facet of every Agile method, and it uses tooling and automation, which some of the Agile community will tell you that you don't need. There is a bit of a double standard—tooling for testing and integration is okay, but tooling for project management is, to some people, unacceptable*—apparently it should all be on index cards and white boards.

•••••••••••••••••••

*At the time, some Agile thought leaders, such as Ken Schwaber, were openly against the use of software tracking tools, such as Rally Development or VersionOne. The Manifesto had clearly stated the movement to hold a preference for "individuals and interactions over processes and tools." Some leaders interpreted this preference very strongly.

Meanwhile, treating people as people isn't new either. It's been 35 years since Jerry Weinberg published *The Psychology of Computer Programming* and 30 years since Fred Brooks's original "Mythical Man Month" article. So, as much as the work of Alistair Cockburn, and others, has highlighted the importance of treating people as people, and realizing that psychology is the big factor in motivation and productivity, I don't feel that the people factor uniquely differentiates Agile software development.

The factor that I believe is unique, compared to earlier approaches, is TRUST.** Trust is the grease that takes the friction out of the software-engineering economy. It's well documented that economies with high levels of trust generate the highest levels of growth and the greatest wealth for their participants. For example, two to three hundred years ago, the trust among Dutch merchants led to the greatest levels of trade going through cities like Amsterdam. This brought great wealth to The Netherlands, and established the Dutch guilder as the world's reserve currency. Throughout history there has been a strong correlation between the highest-trust societies and the reserve currency. Although there was some literature about the importance of trust in software engineering—Luke Homann's *Journey of the Software Professional* comes to mind—it wasn't a recognized aspect of the software-engineering economy until Agile methods put it front and center.

• • • • • • • • • • • • • • • • •

** This is really one of the first explicit acknowledgements that a novel aspect of Agile methods was the inclusion of sociology and an understanding that the interactions between people were a vital part of successful knowledge work. While psychology had been recognized as important since the late 1960's, those searching for the novelty in Agile should have been focusing on sociology. A considerable portion of this text has been dedicated to sociological concerns and management guidance related to the sociology of teams and larger organizations.

• • • • • • • • • • • • • • • • •

The following section reflects directly on the work I was doing for Microsoft in 2005. Working with Granville (Randy) Miller, together we created two process definitions for the Visual Studio Team System product line for use by .Net developers the world over. The first was intended to include the essence of Agile methods while still making Agile palatable for large corporate adoption. In 2005, broad acceptance of Agile methods had not been established in larger companies. The second process definition was intended to provide coverage of CMMI Model Level

3 straight out of the box, while providing many of the benefits of Agile methods. Edits have been made to this piece to keep it relevant for an audience in the second decade of the twenty-first century.

In developing MSF v4.0 in both its Agile Software Development and CMMI Process Improvement flavors, we (at Microsoft) focused on the idea that high levels of trust were important for high productivity. We focused on eliminating the waste in software engineering that comes from lack of trust. (With the CMMI method, we still need to facilitate audits and appraisals, but we've tried to automate as much of the evidence gathering as is possible using the tool within the specification.) *Audits and arbitration are aspects of a low-trust society.* Agile software development brought the idea of trust to the forefront. When there is trust, there is less waste, less extra work, less verification, less auditing, less paperwork, fewer meetings, less finger pointing, and less blame storming. Building trust between the software engineering team and its customers is the first goal for any manager using Agile. Equally, building trust within and among the engineering team is also essential. So many aspects of Agile methods are about building trust: frequent delivery; focus on working code rather than on documentation; face-to-face communication; pair programming; peer reviewing; stand-up meetings; shared responsibility and joint accountability; direct customer collaboration, including on-site customers and customer involvement in modeling, analysis, and estimation; tracking and reporting based on customer-valued functionality; information radiators; big, visible charts; and, ultimately, complete transparency into the entire engineering process. Agile, for the first time, enables us to run software development like other parts of a business. It clears away the fog. It lifts the veil of secrecy. It blows away the opaque clouds and reveals the naked truth of what is really going on.

During the development of MSF v4.0, we faced criticism from some quarters that Microsoft was misusing the "Agile" term, or misappropriating it, or seeking to reinvent it in its own image—Simon Evans provides one recent example.

> I feel annoyed that Microsoft have hijacked the Agile term, and turned into formal. There is nothing Agile about MSF. [1]

Nothing could be further from the truth. What we have done is create an Agile method that addresses previously unsolved problems for typical IT departments. It's a full-lifecycle Agile method that gives responsibilities to a wider range of roles—roles typically found in IT departments. And with our CMMI method, we have stuck closely to our core Agile process definition and stretched it to fit.

1. http://blogs.conchango.com/simonevans/archive/2005/06/09/1586.aspx

Agile Litmus Test

Saturday, February 4, 2006

Documentation, contracts, reviews, memos, other non-verbal communication, verifications, and validations are a reflection on the amount of trust involved.

I ATTENDED A SEMINAR IN BELLEVUE LAST THURSDAY. I GOT TALKING TO LANCE Young over breakfast. I know Lance through the Seattle Java Users Group. He told me he was looking to create an Agile litmus test, and he was building a number of dimensions to the problem and which questions need to be answered, such as, "How heavy is the documentation?" and so forth. I replied that I thought there was only one essential question:

How much trust is there in the room?

If the team is not all in a room together, that is a bad place to start. Assuming they are all in the same room, the more trust there is, the fewer transaction costs there will be. Documentation, contracts, reviews, memos, other non-verbal communication, verifications, and validations are a reflection on the amount of trust involved. These represent the transaction costs of creating great software. Less trust equals more overhead. Less overhead for transaction costs means less waste. Less waste means a better economic outcome. And a better economic outcome should ultimately be the reason we adopt Agile.

Instilling Discipline

October 30, 2003

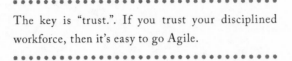

> The key is "trust.". If you trust your disciplined workforce, then it's easy to go Agile.

DISCIPLINE IS OFTEN CITED BY LEADERS IN THE AGILE community as an essential element in the adoption of an Agile method and a transition to a more agile way of working. Writing in *Good to Great*, Jim Collins highlights that companies that made the transition from merely "good" to "great" have a culture of discipline.

> **A Culture of Discipline**. All companies have a culture, some companies have discipline, but few companies have a culture of discipline. When you have disciplined people, you don't need hierarchy. When you have disciplined thought, you don't need bureaucracy. When you have disciplined action, you don't need excessive controls.[1]

I believe this helps us understand an aspect of how twenty-first-century Agile methods differ from their heavyweight twentieth-century forebears. Traditional heavyweight methods replace professional discipline with process in order to verify and enforce rigor, precision, and quality. These methods are heavyweight precisely because they *do not trust* the discipline of the engineers. The key is "trust." If you *trust* your *disciplined* workforce, it's easy to *go Agile*.

Hence, it seems that in order to go Agile, a manager must first start by instilling a culture of discipline in the organization. This seems a necessary precursor for success. So necessary, in fact, that it should probably precede a move to an Agile development method by several months. By first encouraging discipline, the organization will make a smoother, faster transition to Agile methods, and the results will be better sooner and more easily sustained later.

---

1. *Jim Collins, "Good to Great" p.13 [refs to be fixed later]*

Looking back on this eight years later, it seems rather speculative—and probably wrong. I no longer believe that a manager must focus on discipline prior to introducing an Agile method. However, I do believe that a focus on discipline must be part of it. I feel now, writing in 2012, that discipline at both the individual and the team level is a necessary condition for successful Agile adoption.

It is ironic, then, that Barry Boehm and Richard Turner chose to counter the term "agility" with the term "discipline," in their book, *Balancing Agility and Discipline–A Guide for the Perplexed* (Addison-Wesley, 2003). It's often misunderstood that Boehm and Turner were proposing that traditional methods of software development were disciplined, but that Agile methods were not. In fact, they were proposing a risk profile for projects that exhibited a spectrum where "agility" (note the small a) was at one end and "discipline" at the other. Their work then suggests that selecting an Agile method should be matched to an appropriate risk profile. Such work on risk was uncommon in the Agile community at the time.

If we are looking for a term to juxtapose with Agile, I much prefer "rigor," coined by Jim Highsmith in *Agile Software Development Ecosystems*. Highsmith describes traditional methods as "rigorous software methods" (or RSMs). The term "rigor" describes the situation more accurately, in my opinion. Rigor does not imply trust or discipline, but it does imply attention to detail and quality. Traditional methods are designed to deliver quality through attention to detail. They don't assume any standard of discipline.

Making Progress with
Imperfect Information

Tuesday February 27, 2007

• •

While a true differentiator of Agile methods was
is their embracing of a high- trust culture—with all
the benefits that delivers—the true essence of Agile
wais the concept of moving forward with imperfect
information and being prepared to rework later.

• •

MY DEPTH OF UNDERSTANDING OF AGILE VALUES, PRINCIPLES, AND MOTIVATIONS continued to grow through the middle part of the last decade. While I'd identified high levels of social capital, or trust, as perhaps the true differentiator of the Agile era, I came to realize that there were some other dimensions. The one that had been hiding in plain sight, as a practice, became evident to me as a principle around the winter of 2007.

I've been re-evaluating my view on refactoring!

It's interesting how your vantage point can color your view of something. My perspective on the Agile community and its practices has always been rooted in my experience with Feature-Driven Development and the Coad Method. When I spoke at USC[1] back in 2004, I suggested that although Feature-Driven Development might never be seen as the most agile of Agile methods, it was probably the most Lean. I stand by this comment three years later. FDD is very Lean. All the waste is trimmed out. The Coad Method techniques deliver a precise definition of a domain that is loosely coupled and highly cohesive, while the Feature definition technique delivers fine-grained customer-valued units of work that exhibit very low degrees of variation in size and complexity. The batching technique of Chief Programmer Work Packages is very efficient and minimizes transaction costs associated with work-in-process. I could go on, but you get the point—FDD is Lean. The modeling, planning, and batching for design and build result in almost no need to refactor code on FDD projects. As a

---

1. http://www.agilemanagement.net/index.php/blog/USC_Annual_Research_Review

result, my book, *Agile Management for Software Engineering,* classified refactoring as rework, and labeled it as waste.

From my FDD vantage point, that might have been a fair assessment. However, now that I'm running a software-engineering organization, with which I inherited a waterfall process that runs through a series of narrow and specialized departments such as Business Analysis, Systems Analysis, and Development and Test, I've changed my opinion. From a new perspective, refactoring is clearly a very valuable process.

I believe that Alistair Cockburn's paper from the ICAM 2005 International Conference on Agility[1] will come to be seen as a seminal paper in Lean Software Engineering. In this paper, Cockburn explains that asking a non-bottleneck resource to do extra work to rework something does not cost anything extra, and it can create a desirable effect because it allows progress to be made and demonstrated sooner. Ergo, rework is not waste when performed by a non-bottleneck resource.

There have been many versions of the idea that perfect is the enemy of good enough. Its origins are perhaps traced to Voltaire, who said, "The best is the enemy of the good." And it is this concept that refactoring (and Cockburn's paper) embrace. He argues that it is better to make progress with imperfect information, and refactor later when better information is available, than to wait for better information before progressing.

Specifically, I am thinking that it is better for developers to start coding with imperfect analysis than to wait for a systems analyst to produce a "perfect" specification.* The developer can then refactor the code when the analyst makes a final version of the specification available. My reasoning is simple. The developer would otherwise be idle. (Not truly idle, as there is plenty of busy work available—grooming environments, training on new languages and APIs, and so forth, but idle in the sense that he or she is not adding value to the deliverable.) By definition, a resource (or station) with idle time is a not a bottleneck resource. It is, therefore, okay to aska developer to perform refactoring. The refactoring cannot be classified as waste in a case like this.

We can think about this decision using real-option analysis. The option we are buying is to deliver the working code sooner. The cost of the option is the cost of having the developer start work before a final specification is ready. The risk (or uncertainty) attached to the option is the risk that the early, imperfect specification will be significantly different from the final specification, and

1. Cockburn, Alistair, "Two Case Studies Motivating Efficiency as a 'Spendable' Quantity" http://alistair.cockburn.us/Two+Case+Studies+Motivating+Efficiency+as+a+%22Spendable%22+Quantity/v/slim

that any rework will take longer than waiting to start coding on delivery of the final specification. Note that the rework may absorb all of the slack in the non-bottleneck resource, turning it into a bottleneck and delaying the whole project. This gives us a framework to decide whether starting early and refactoring is the correct decision, or whether waiting and coding for "right the first time" is the correct decision.

* * * * * * * * * * * * * * * * *

* I've come to realize that this notion is at the heart of eXtreme Programming, and it was described in Kent Beck's seminal book on the topic, *Extreme Programming Explained–Embrace Change* (Addison-Wesley, 2e, 2004). Beck refuted, though without any truly scientific study, Boehm's finding from 1976 on cost of change. Boehm's work and subsequent corroborating evidence suggested that late-breaking changes cost exponentially more. The Boehm evidence on the economics of software development favored the pursuit of perfect information early in the lifecycle. This ultimately led to the development of more and more analysis and design techniques that pursued greater and greater levels of information discovery earlier in the lifecycle. The core mistake is to assume that this is even possible. The nature of software development is too complex for this to work. What is required is to probe for more information with partial or approximate solutions, and implement a feedback loop so as to iteratively pursue the answer. Boehm's later work on Spiral Model in the 1980s showed that he understood this. The rest of the industry did not. Ultimately, the pressure from heavyweight processes, laden with analysis and design techniques, was to provoke the Agile revolution in the late 1990s. So, while a true differentiator of Agile methods is their embracing of a high-trust culture–with all the benefits that delivers–the true essence of Agile is the concept of moving forward with imperfect information and being prepared to rework later. Rework wasn't waste in the traditional twentieth-century quality-assurance movement's understanding; rather, it was further information discovery based on feedback.

Spiral Model Revisited

Friday August 6, 2004

• •

Boehm described Spiral as evolutionary rather
than incremental. The purpose of an evolutionary
approach was better risk management, enabled by an
ability to analyze the whole system

• •

I FEEL MORE AND MORE THAT WHAT I AND OTHERS ARE DOING WITH AGILE
comes back to Boehm's Spiral Model first postulated in 1986,[1] and refined in
1988.[2] Boehm described Spiral as *evolutionary* rather than incremental.* The
purpose of an evolutionary approach was better risk management, enabled by
an ability to analyze the whole system, but only in the briefest detail, to identify
the highest-priority features for an initial iteration. On delivery of the iteration,
customer feedback would be used to prioritize the next set of features, make
refinements, and add detail to the initial analysis. It was this feedback and
refinement that separated it from merely incremental development—presumably
the assumption in 1986 was that incremental meant big analysis up-front, divide
the system into set of iterations, and then perform each iteration in open loop
without accepting changes.

I've been reflecting on a big project I was involved in during the first half of
the year; it started its discovery phase in the fourth quarter of last year. Without
realizing it was significant at the time, I now reflect on the fact that we did two
passes of modeling—this is not part of the standard Feature-Driven Development
prescription. The domain was so new that no one really knew what was wanted
or what the system should do. We did a first-pass domain modeling session with
all the marketing and project management folks, as well as with the architects
and chief programmers. This initial pass was very thin—because we knew so
little about the domain. However, this served its purpose to flush out areas for
further discussion, and as an enabler to allow the marketing folks to define their

---

1. Barry Boehm, "A Spiral Model of Software Development and Enhancement", *ACM SIGSOFT Software Engineering Notes*, August 1986.

2. Barry Boehm "A Spiral Model of Software Development and Enhancement", *IEEE Computer*, vol.21, #5, May 1988, pp 61-72.

requirements. This first-pass modeling session gave us a scope, a list of business processes, and an outline of the subject areas in the domain. It was enough for us to identify rough outlines for six iterations.

The second pass of modeling—in greater detail, but no greater than a typical FDD project—then happened on each increment of development, and typically took four to five days. The increments were time-boxed to seven weeks. This delivery cadence was set to enable system integration testing with other components. One week of modeling for a total seven-week increment of development—this is a lot more time spent modeling than Jeff De Luca or Peter Coad would typically recommend (one week for every three months is the norm), but the domain was new and uncertain, which probably explains the extra effort required. A Feature List was written, and the iteration scoped in detail. The Design-By-Feature, Build-by-Feature cycle then started, and each Feature was designed with a UML sequence diagram. Hence, there were three passes at modeling, and on each pass, more information was obtained. This is one more than a typical FDD project, and it begins to sound more and more like Boehm's Spiral Model.

When I first heard about Spiral Model, I was dismissive. The key to it is an ability to examine a problem in only enough detail to move around the loop. It's been widely identified that over-modeling,[1] or drilling deep, is a common problem with developers. The natural tendency is toward "big design up-front." What makes the Spiral Model hard to achieve in practice is the discipline, or craft awareness, that enables a wide-but-not-deep pass of the entire scope in order to manage risk and prioritize deliverables. At the time, I didn't believe that a suitable technique—a suitable enabler—existed to make Spiral Model real in practice. However, I am now wondering whether Coad's domain modeling with color archetypes is such a tool, an enabler of a true spiral approach to software engineering.

I ask myself, would I do it again—domain model in two passes (or iterations) for a multi-increment, large project? The answer is definitely, "Yes!" Do I think that developers can be taught to model just wide enough to embrace uncertainty, and to understand when they have just enough information to move forward? The answer to this is also "Yes!" but there is a caveat. This skill—of knowing when to stop—is learned through experience. Hence, the skill can only be passed from individual to individual through real-world work experience and direct observation. It truly is a craft. Could we (the experts in domain modeling in color) codify that experience so that it might be taught in a classroom? I don't have an answer for that yet.**

1. http://www.agilemanagement.net/index.php/blog/Quality_Assurance_Over-modeling/

* It is worth noting that the Kanban Method introduced an evolutionary approach to change—process evolution—and Lean Startup takes an evolutionary approach to product or service development. All of a sudden "evolutionary" is the great new idea. Ironic then that Boehm was arguing this as early as 1988—long before the Agile revolution. Boehm's assessment that we need an evolutionary approach to software development was both inspired, given the timing (the mid-1980s), and correct. eXtreme Programming that inspired various flavors of test-driven development has given us an evolutionary approach to software development. Kanban has given us an evolutionary approach to adapting software development lifecycle processes. Lean Startup has given us an evolutionary approach to product and service definition. As a profession and an industry, we are learning how to create evolutionary methods.

** Another valuable lesson from this article is that there is more than one way to acquire (partial) information. Although test-driven development has been the dominant popular approach in the twenty-first century, it is evident that older modeling and analysis methods can be used with equal effect, in the hands of skilled craftsmen. What is important is that the output can be used to probe for further information and to provide useful feedback. In this sense, it is undoubtedly true that working software produces better results with a broader audience – a demonstrable piece of software is more useful as an information probe than a UML class diagram of a domain. Viewed through the lens of complexity science, it is easy to see why the legacy of eXtreme Programming is so valuable.

What Are the Right Conditions
for Agile Adoption?

Monday September 1, 2008

. . . .

If Agile is all about high- trust culture, and Holland
is one of the highest-trust countries in the world,
why is Agile adoption in Holland so slow?

. . . .

For several years, I've been promoting the notion that "Trust is the essence of Agile." This year, I've been articulating that the statement "[We value] Customer Collaboration over Contract Negotiation" is the statement in the Agile Manifesto that really captures this concept. High-trust, high social-capital cultures allow the overhead of contract negotiation, commitments, audit, and arbitration to be eliminated—which saves time and money.

Back in December 2007, I was in Belgium for the Javapolis event, giving an evening talk to the Belgium XP Users Group. Some of the Dutch attendees came up to me and said, "So, if Agile is all about high-trust culture, and Holland is one of the highest-trust countries in the world, why is Agile adoption in Holland so slow?" This got me thinking.

The answer came to me early this year, while I was reading *Collapse* (Penguin, 2005) by Jared Diamond. This is a book many claim to have read. It is a best seller; you can get it in many airport bookstalls. But it is one of the densest and most detailed texts I've ever undertaken. It took me about four months to read it through, but it was worth the effort. I would suggest that the most valuable and relevant insight for us, the Agile community, comes in Chapter 9—the discussion of the demise of the Greenland Norse.

The Norse were actually in Greenland before the Inuit. They were eventually displaced by the Inuit, who were moving south as Earth cooled in the early part of the second millennium a.d. The climate change eradicated the Greenland Norse. The Inuit were not native to the region. They had moved in and displaced the Dorset people to the north some centuries earlier. They rarely encountered the Norse, but when they did, it led to fighting. Eventually, the Inuit eliminated most of the Norse population in a battle one summer. The rest of the population appears to have perished in the winter of the following year. So, what went wrong?

It seems the Greenland Norse would not change and would not adapt to changing conditions around them. They remained European. They practiced European farming methods. They had a European structure to their society, including a cathedral, a bishop, and a tribal hierarchy associated with their government. They took their cues from these civic and religious leaders. It seems they stopped eating fish early in their tenure in Greenland. Why? No one knows. They maintained a mainly European diet, consisting of animals they raised and crops they farmed. As the climate cooled, the crop yields from the land would not sustain them or the animals, and they eventually starved to death.

Meanwhile, the Inuit thrived. The Norse had the opportunity to observe the Inuit. There is written evidence. In fact, it appears that every Norse encounter with the Inuit was documented. It seems the Norse believed themselves superior to the Inuit, who were savages and heathens because they were not of the Christian (Roman Catholic) faith. Hence, the Norse did not value the Inuit or their ways. But the Inuit were thriving. They had fast canoes. They could fish. They ate fish. And they had survival techniques that allowed them to cope with changing conditions. But as the Norse slowly perished, they did nothing to help themselves. They refused to change.

To (perhaps overly) simplify, the Greenland Norse were a conservative culture. They stuck to what they knew. They continued to do things the way they always had, and they hoped it would see them through adversity. In the end, this conservative approach, which was resistant to change and preferred the status quo, is what killed them. Had they been a liberal culture, ready to embrace new methods—and open to the influence of other people and their ideas—the Norse may well have adapted and eventually assimilated with the Inuit, and lived harmoniously with them. But they didn't. Refusal to adapt to change killed them. They ceased to exist in the fourteenth century.

This brought me to the conclusion that the ideal circumstances for Agile to thrive and to gain adoption is in cultures that have both high social capital (high levels of trust), and are liberal (small "l") in thinking—open to new ideas, new thinking, and influence from outsiders. Although Holland is seen as a liberal country in several social ways—soft drugs are legalized (regulated and taxed), and so is prostitution—in business, Holland is a conservative country. Any Dutch people I've discussed this with have tended to nod vigorously in agreement.

The other very high-trust culture where Agile has struggled for adoption is Japan. Again, despite the evidence of baby goths in Harijuku on a Sunday—which suggests a socially liberal culture—in business, Japan is truly a conservative nation.

So where is the broadest Agile adoption in Europe? It's mainly in Britain and the Nordic countries. The Scandinavian countries (Norway, Denmark, Sweden) are all high-trust cultures, though not (I believe) quite as high as Holland or Japan (or perhaps it is the length of track record that matters—Holland and Japan have been recognized as high–social capital countries for as long as 300+ years). So, why is Agile adoption in the Nordic region more prevalent? Is it because businesses in Finland, Sweden, Norway, and Denmark are more open to change? Are they more open to outside ideas? More open to influences? Are they generally more adaptable in their outlook to process and business methods? (I'm asking the question. The Scandinavians and Finns reading this can tell me if they agree.) As for Britain, it's a pretty liberal culture, despite its reputation. It isn't a particularly high-trust culture, but neither is it a low-trust culture.

Meanwhile, the low rate of Agile adoption in southern Europe and the Latin countries does seem to be predicted by a lack of social capital in society and/or a conservative approach to business.

What I'm wondering now is whether Agile can become established in low–social capital, conservative cultures. Is it simply that it takes time, or will it never be adopted? This begs the question: Will the economies of these countries suffer as they fail to adapt to change? Will they suffer an economically similar fate to the Greenland Norse? As the business climate changes and the twenty-first century unfolds, will they fail to adapt, and, instead of being eliminated, simply become second- or third-rate—with greatly reduced relative economic performance and a significant drop in GDP per capita, ranked against other countries?

It will be interesting to watch.

• • • • • • • • • • • • • • • • •

*Writing in 2012, we now know that the Kanban Method has been adopted much more widely, in a shorter period of time, than Agile software development methods were a decade earlier. While early Kanban adoption followed that of early Agile development adoption, it quickly spread into countries like the Netherlands, Switzerland, Brazil, and South Africa, with pockets of adoption in France, Spain, Portugal, Chile, Argentina, Peru, and even China. The question is, why? Michael Sahota has suggested that Kanban is better aligned with a "control culture," as defined by William E. Schneider. Schneider's model divides corporate culture into four categories: control, collaboration, cultivation, and competence. Sahota observes that Agile development is best aligned with collaboration, cultivation, and competence, and not at all with control. However, most low-trust, conservative business cultures have a control mindset. Hence, Schneider's control culture is actually dominant in business, and it is certainly prevalent in lower-trust, conservative countries. This likely offers us an explanation for the more rapid adoption of Kanban in these countries and among larger business. Kanban is a method better suited to adapting and evolving large corporations with a conservative, low-trust, controlling approach to management, corporate governance, and investment.[1]

---

1. References: *Trust* by Francis Fukuyama, *Collapse* by Jared DiamondMichael Sahota, "Kanban Aligns with Control Culture" http://agilitrix.com/2011/04/kanban-aligns-with-control-culture/

Extreme Brandy Butter

Tuesday, December 27, 2005

Traditions are important. Family traditions, held as tacit knowledge, passed from father to son, equally so. This Christmas Day I reprised an Anderson family tradition—making the brandy butter to be used later with Christmas pudding and mince pies as part of our traditional British Christmas feast. The Anderson family takes their dessert after the Christmas turkey very seriously, indeed. We're extreme brandy-butter makers.

Having a cancer patient in the family is difficult enough, but when it is known to be terminal and the patient is a nine-hour flight by 747 away from insurance-covered health care, it is all the more difficult. However, two years ago, my father was given the opportunity to spend Christmas with us, here in Seattle. My mother had died that past September, and for the first time since a business trip to Frankfurt, in 1975, he was able to travel abroad. My mother wasn't a traveler. Her idea of a vacation was four days at a bed and breakfast within three hours' drive from home. She just didn't like to be away. Nor did she like to be alone.

It's an ill wind that doesn't blow some good, and that year the good that blew was the opportunity to have three generations of my family together, here in Seattle. Both my father and I were well aware that it might be his last. We were determined to make it work, despite the health difficulties, and equally, we were determined to make it fun. It was therefore necessary to pass on the tradition of making the brandy butter on Christmas Day to me—the eldest son. It seems strange that for all those years I had enjoyed the eating, but I had never inquired about the making, nor had I ventured to experiment with it. It was implicitly the job of the head of household. So with my dad looking over my shoulder (do we call this pairing?), I prepared the ultimate condiment for the pudding. After all, it is basically fat, sugar, and alcohol; what more could a body need to be happy?

To follow along you will need (at least) eight ounces of unsalted butter—the kind of butter that, if it consulted with a hypnotist, could still regress back to a memory of cows in a green pasture from its former life as milk. You'll need a bag of fine-grained baker's sugar (confectioners' sugar, to some). No need for measurements—this is extreme brandy-butter making. For those who know anyone in the food preparation trade, you'll be familiar with the idea that cooking is an art, while baking is a science. Baking is chemistry. It is all about precise measurements, and precise techniques of preparation—folding, not beating, for example. In theory, brandy-butter making is chemistry. I'm sure it is possible to create the ultimate six sigma brandy butter. Indeed, the manufacturer that produces the small jars that can be had from Cost Plus World Market in downtown Seattle for seven bucks a pop must surely have some chemistry and some quality assurance on their side. However, for the Andersons, making brandy butter is a craft—indeed, an art form. So, just a bag of sugar will do fine. You'll also need some brandy. Again, no measurements required—any old bottle of very special (VS) port or Cognac will do. You'll need a small baking bowl, a knife, a fork (not for the extreme b-b maker any modern-fangled tools like whisks or hand-blenders; we prefer individuals and their interactions with the ingredients over processes and tools), and a spatula.

Cut the butter into chunks with the knife, about 4 to 8 chunks will do—no need to be precise. We're going to practice ECUF (enough creaming up-front) also known as JECI (just enough creaming initially), so add all the chunks to the bowl. Open the bag of sugar and shake a handsome portion onto the butter. It should probably cover the butter cubes, but not drown them. Squeeze down on the butter gently, and stir with the fork to slowly work the sugar into the butter. Gradually, it will cream together and become a smooth, single mass. At this point, repeat pouring the sugar, and mix as vigorously as possible with the fork. It's now time to sample the first feedback. Use the spatula and take a small sample from the bowl. Sweet enough? Does it taste like mildly buttered sugar? If not, repeat the sugar pouring and beating and frequently sample some with the spatula for more feedback. When the creamed sugar is prepared to your satisfaction, you are ready for the highly iterative process of fortifying it to create the finished product.

Plan on a large first iteration. Liberally, pour the brandy into the bowl—no sixth-gill measures here. Think well left of Bill Clinton for your definition of liberal (for those reading from the archipelago just northwest of Europe, insert Charles Kennedy or your favorite soft politician instead.) Gradually stir the brandy into the creamed sugar. It will begin to turn a champagne-brown color. When the brandy is completely absorbed, use the spatula to sample some feedback. The second iteration can probably be of a similar length. Be generous; tell the Cognac bottle it is better to give than to receive. Mix it in again. It will take a little longer to be fully absorbed. The color will darken slightly. When sampling the feedback this time around, you should be able to detect the brandy flavor. From this point on, smaller iterations will be required. More feedback sampling will be needed to blend the perfect batch of brandy butter.

"How do you know when you're done, Dad?"

"When you've added as much brandy as you can, without it separating later into its constituent parts, then you're done! You remember that year when I overdid it slightly?"

"Yes. It separated in the fridge. Physics. The lightest stuff floats to the top. Beautiful set of layers."

"Well, I learned my lesson."

So, small iterations, rapid feedback. Obtaining that perfect mixture for ultimate taste (and effect) needs an empirical process with oft-sampled feedback on the output.

Add a thimbleful (approximately, no thimbles are actually used) of brandy. Stir. Taste. Again. Stir. Taste. At a point very soon, the creamed butter will come loose from the bowl and slide around on a thin ice of brandy. To the uninitiated, it looks like you're done. Wrong! You are about halfway there. Stir more vigorously. The brandy will be absorbed. Repeat. Another thimble. Stir vigorously. Taste. Repeat. Again. And yet again. Are you getting a taste for this?

If, suddenly, you realize that maybe, just maybe, you don't have enough product remaining in the bowl, you are a craftsman, an artist, a scholar of the brandy-butter creaming process. You'll need to make a second bowl later. Don't stop to get more butter from the fridge. If, on the other hand, you find that you've had to make a mad dash to the drinks cabinet in the sitting room to raid the five-star VSOP in order to provide raw material for future iterations, I salute you!

By this time, your eyes should be watering from the fumes rising from the bowl. If they are not, you might like to rethink your regular weekly alcohol consumption.

How does it taste now? Does it skid around the bowl for tens of seconds before absorption of each thimblefull of brandy? Is your feedback interpretation, "Pretty darn good, potent stuff"? If so, then just one more iteration should do it. Just for luck!

Decant the remaining finished product into a serving dish, cover, and place in the fridge until after dinner. It should harden up just enough to be pleasant to cut and serve, and should then melt into the pudding within a minute. Eat quickly for maximum pleasure.

Agile Practices

I'VE SPENT THE AGILE ERA AS A MANAGER, CONSULTANT, TRAINER AND THOUGHT LEADER. I HAVEN'T BEEN A PRACTITIONER IN SOFTWARE DEVELOPMENT IN THE TWENTY-FIRST CENTURY. SO I SELDOM BLOGGED THOUGHTS ON AGILE PRACTICES. HOWEVER, A RECURRING THEME FOR ME HAS been the practices surrounding planning and the inputs for such plans—estimates.

I often see Agile practices go unchallenged. Few in the community have been brave enough to challenge the textbook recipes or call out waste when they see it. Some of the braver ones have looked to the Lean principle of identifying and eliminating waste. Arlo Belshee, Jim Shore, and Joshua Kerievsky all refined eXtreme Progamming to give it more of a flow-oriented flavor—more kanban-like. Their motivation for this was reducing waste in XP. As I explained in *Kanban: Successful Evolutionary Change for Your Technology Business*, they had a different reason and motivation for introducing kanban-like practices than I did. They were not pursuing an evolutionary approach to change. However, the resultant processes were remarkably similar.

If the Agile movement is to remain healthy, it must be prepared to challenge established practice, and those who do so should be given a voice, a forum, and an opportunity to show leadership.

For many of the social and tribal reasons you now understand, my challenges to established Agile practices were often interpreted as an attack on Agile rather than an attempt at refinement and improvement. Judge for yourself . . .

Agile Apprenticeship

Monday, May 3, 2004

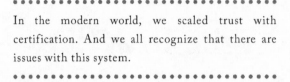

In the modern world, we scaled trust with certification. And we all recognize that there are issues with this system.

MARTIN FOWLER ASKS, "SHOULD THERE BE AGILE CERTIFICATION?" AND speculates that perhaps people wouldn't pay for it. This is a theme that Steve McConnell touched on in his book, *Professional Software Development* (Addison-Wesley, 2003), in which, in Chapter 19, "Stinking Badges," he takes the idea further than certification and talks openly about licensing, ahem, just like real engineers.

> I remember a conversation over beer after XP Universe 2002. We wondered what would be needed for an XP certification. We considered that it would involve several weeks of observation, watching people deal with the various stages of a software project, seeing them use various skills, including tuning the process. Such a test would be expensive to carry out. Would people be prepared to pay for this kind of program?[1]
>
> —Martin Fowler

The real issue with this is economics—the market currently does not demand it. It really requires that firms with qualified (or certified people) visibly beat out competitors through their use of more agile teams—only then will the market start to demand certification or licensing.

You need to compare effectiveness against competency, and see if there is a relationship. Competency can be measured only by some form of certification. The prolonged certification process that Martin Fowler suggests sounds more like an apprenticeship to me. Yes, Agile development is a craft, and perhaps a journeyman scale and a craftsman's certification model is what is required. Initially, this starts small because the number of experts and masters in the craft is small. Hence, the number of journeymen is small, and the number of apprentices is not quite so small. However, gradually the numbers would swell.

1. http://martinfowler.com/bliki/AgileCertification.html

How does a craftsman report his or her status in the profession or craft? Perhaps like this:

> David Anderson trained under Peter Coad and Jeff De Luca. Worked as journeyman at <insert employers> in the following <capacity>. Achieved his master craftsman status after delivering/building <project> or publishing <work>. Finally, he achieved his expert status when invited to write and later publish *Agile Management for Software Engineering* in Peter Coad's series for Prentice Hall—a work that extended the state of the art in software lifecycle management through application of the Theory of Constraints.

The challenge with this is one of scale. Craft guilds are an ancient tradition that died out in the eighteenth and nineteenth centuries; global trade disconnected buyers from artisans through distribution channels involving a series of merchants and brokers. The replacement for the social status conferred by a craft apprenticeship was replaced by certification from bodies of sufficient authority such that their brand carried a level of trust in the market. University degrees are a form of certification that scale globally and carry a trust mark. My degree from the University of Strathclyde in Glasgow, Scotland, is sufficient to enable me to acquire a visa to work in the United States, and eventually to gain permanent residency. The immigration authorities do not know my alma mater or the staff who trained me. They rely on a third-party service that validates degrees by checking the status of the accredited institution and the level of the degree against some loosely defined international standards. In the modern world, we scaled trust with certification. And we all recognize that there are issues with this system.

What would it take to reproduce a craft guild system for the global flat world of the twenty-first century?

Everyone in the industry would have to produce a résumé similar to the example above, one that was somehow rooted in the core experts of Agile (or some other branch of software engineering) as the original mentors. For example, my staff from Sprint in Kansas City would list that they trained under me, and that I, in turn, had trained under Peter Coad and Jeff De Luca. Everyone would have a pedigree. This would need a social-media technology similar in nature to the recent LinkedIn service that is becoming popular. Perhaps social media will provide the answer to twenty-first–century trust in competency, capability, and social status within a profession. However, I can't see this happening anytime soon.

●●●●●●●●●●●●●●●●●●

*Editing and writing this compendium of articles in 2012, I find myself involved in the creation of Lean Kanban University, a standards body and training accreditation organization for Kanban applied to knowledge-work industries such as software development. An active part of our discussion is how we might provide such social-media technology to record the career timeline and history for individual members who want to prove their worth within the community and to future employers. Eight years on since this original blog post, we are still waiting for a viable alternative to certification.

Why We Lost Focus on Development Practices

Wednesday, November 26, 2008

• •

The community tends to get sucked to where the constraining problems are occurring. This is the natural and correct behavior.

• •

FOLLOWING THE LATIN AMERICAN AGILE CONFERENCE IN BUENOS AIRES, Tobias Mayer blogged his reaction to the panel session.[1] He seemed to take issue with some comments from Mary Poppendieck. I'd like to offer my own response to Tobias, specifically to the interaction between him and Micah Martin in the blog comments. Here is an excerpt from Micah's comments:

> . . . My frustration is not specifically with Scrum but with the diminished focus on software in the Agile community. As Agile has grown it has become more of a project management topic rather than a software development topic. For good or bad, Scrum is the most prominent face of Agile project management and so it gets the blame. . . . —Micah Martin

I've heard this sentiment a lot recently at conferences from the programming-focused folks. The key problem is that their locus of interest surrounds programming, and they want to be part of a community inventing *better ways of developing software* (in the tightest definition and meaning of *programming*). Meanwhile, many of us in the Agile community are interested in *better ways of developing software* (projects and products). This second interpretation of the purpose behind the Agile Manifesto leads us to look at whatever constrains our ability to deliver projects and products better, faster, cheaper, and with high quality and high value.

It's understandable that many developers want a focus on their own craft. And it is understandable that without this they quickly lose interest. As Tom DeMarco wrote, they are Dilbert, delegating responsibility for their success

---

1. Mayer, Tobias, "Getting Trashed by the Lean Machine" http://agilethinking.net/blog/2008/10/23/getting-trashed-by-the-lean-machine/

to someone else. It's a "blinders-on and just leave me alone to do my thing," approach.

While I solidly believe that there is a lot more we have to learn about programming and software architecture and engineering, I think the reason that the community has lost its focus on this activity is easily explained. *Programming and programmers are not the constraining factor on improved performance!*

In my class, I often teach participants about bottlenecks. I ask them, in a collaborative exercise, to speculate about the bottleneck in their own organizations and to discuss how they would prove it and what they would do about it. In almost three years of teaching this class, on five continents, and on about 12 occasions—with groups ranging from 16 people to 250 people (at Javapolis in 2007)—I have concluded that developers are the constraining factor on projects and team performance less than ten percent of the time. In some groups it is as low as three percent.

My belief around this is quite simple. Basic Agile practices that focus on quality—including collaborative working, such as pair programming—and have a strong focus on unit testing and early defect detection using continuous integration and test automation—have greatly improved software development, to the point where initial software quality (that is, bug insertion rates) is not the constraining factor on team performance. With immature teams having sloppy development practices and poor initial quality (that is, high defect-insertion rates), development is the constraint. Agile has successfully fixed this!

So, we can declare victory. In this sense, the crowd who argue for a continued focus on better development practices are fighting the last war.

The rest of the community has moved its focus to other areas—most notably project management and business analysis/value-based software engineering. The community tends to get sucked to where the constraining problems are occurring. This is the natural and correct behavior. And it shows that many in the community interpret "better ways of developing software" in the broad sense, rather than the narrow, programming-centric sense.

So, what next for the purely coding-centric minds in the community? I would encourage them to keep at it. As the wider community fixes the issues with project management, analysis, and requirements, and improves the flow of value, the focus will swing back to engineering and development practices. In the case study from Microsoft's XIT business unit,[1] described in chapter 4 of my book, *Kanban: Successful Evolutionary Change for Your Technology Business*, I show how we improved the productivity of a team more than three-fold and

---

1. http://www.agilemanagement.net/index.php/blog/TOC_ICO_Conference_-_Miami_Oct_25

shortened their lead time by 90 percent. All of this was achieved without making any changes to how software was developed or tested. We achieved this by focusing on the bottlenecks and the waste—and eliminating them. What would it have taken to further improve the performance of that team? A renewed focus on development and testing practices!

I truly believe the community will swing back to the developers and the testers. Meanwhile, they need to be patient. It's a compliment not to be the focus of attention. It reflects success. Embrace that success and stop complaining!

Agile Estimating!

Thursday, March 17, 2005

• •

This is lightweight estimating! It's estimating using a model and an existing data set to develop a probabilistic view of the future.

• •

IN CHAPTER 5, I SUGGESTED THAT YOU SHOULD STOP ESTIMATING. I HOPE IT got your attention.

By asking you to stop estimating, I'm suggesting that traditional forms of estimation—take a list of tasks and ask for a level of effort for each one (in man-hours or days)—is a waste of capability. It's better to have some form of inventory measure. A fine-grained one based on some quick, broad, and cheap analysis would be ideal—like Features in Feature-Driven Development—but anything else is just fine, too, for example, Change Requests, Bugs, Scenarios, Use Cases, User Stories. Now, don't estimate them, but instead measure the velocity—how many you can do in a given time—what is the throughput rate? And what is the spread of variation in those numbers?* Use this to estimate the mean time to complete a given batch of inventory and the required buffer for variation.

This is lightweight estimating! It's estimating using a model and an existing data set to develop a probabilistic view of the future—assess the number of value items, multiple by the mean velocity, and buffer for variation. An agile estimate should have a number of value units, a target mean velocity, an end date, and a buffer (or a measure of variation in velocity). That's it. You shouldn't be estimating anything individually. Fine-grained individual estimates of effort are waste—*muda*! Just say, "No!"

• • • • • • • • • • • • • • • • •

*Evidence gathered since writing this post eight years ago has shown that velocity or completion rate variability can be quite excessive-typically in the range of 0.5 to 2.0 times the mean number. As a result of this, I no longer recommend using velocity data for short-term planning.

Why Estimates Are *Muda*!

Tuesday, September 27, 2005

• •

When we spend time estimating something, and
we use the capacity-constrained resources to do the
estimating, we effectively lower our capability.

• •

I'M GETTING A REPUTATION AS A *BAD BOY* OF PROJECT MANAGEMENT! MY ADVICE
to stop estimating doesn't go down too well with the traditional PM community.
It doesn't sit too well with some of my new friends in the traditional software-
engineering community either—their Personal Software Process/Team Software
Process (PSP/TSP) method is based around estimating and then tracking
estimates against actuals. My dislike for earned value reporting isn't too popular
either—particularly as the American government mandates it for certain types
of contracts. (I'm not overly fond of Scrum burndown charts either—when they
are based on time-on-task estimates rather than customer-valued work items
like user stories.) My agile estimating technique based on the velocity of client-
valued work items seems natural to me. It seems like the Agile way. The simple,
easy-to-calculate way. The doesn't-waste-any-time way. And this is what I want
to talk about today—doesn't-waste-any-time! Traditional estimating is *muda*!
Agile estimating isn't! Here is why:

Let's assume that software development is the capacity-constrained resource
in our organization. Even if it isn't, we wouldn't want to waste slack capacity that
might otherwise be used to absorb variation elsewhere in our system. When
we spend time estimating something, and we use the capacity-constrained resources to do the estimating, we effectively lower our capability. We lower our
capability on an activity that isn't client valued. The customer doesn't care how
long your estimate for a task was. Doing the estimate often takes a significant
chunk of time compared to the time it would take to do the work. Doing the
estimate even a few days (but more likely a few weeks or months, and, in some
cases, years) before the work is done means that anything you learned from the
estimating process is lost. It gets worse. Often we estimate work that gets cut or
doesn't get done at all because the estimate is too large. Calculating a time-on-
task estimate doesn't create customer-valued knowledge. Estimating is non-value
added. Estimating is *muda*!

On the other hand, I thoroughly embrace the idea that we analyze and partition our problem space. In Feature-Driven Development, we analyze the work into a domain model, and later we partition it into components. We further analyze the work into Features and group them into Feature Sets and Subject Areas. All of this analysis work gives us a work-breakdown structure that is entirely value-added. All the work done analyzing for the Feature Plan is value-added. It creates knowledge that is used to deliver the customer-valued functionality. We then estimate based on feature velocity—an agile estimating technique that takes almost no effort to calculate. Agile estimating based on analysis minimizes waste of capacity in the capacity-constrained knowledge-worker resources.

So, to be clear—I am all for analysis that produces knowledge we can use in the customer delivery. I am against estimation that produces information that is not of deliverable value to the customer!

Analysis rather than Estimation

Sunday, December 18, 2005

• •

Analysis adds value by creating knowledge about what components we need to create and how they integrate together to synthesize a whole solution. Estimation merely quantifies the cost.

• •

I'VE NOTICED SOME CONFUSION OVER THE DIFFERENCE BETWEEN ESTIMATION and analysis. When I've suggested that we stop estimating and focus on analysis, some people have struggled to understand the difference. Again, when I suggested that all estimates are *muda*, but continued to encourage the use of analysis, some people struggled to understand the difference. So, I decided to look up the dictionary meanings for clarification.

Verbs first:

TO ESTIMATE: to calculate approximately (the amount, extent, magnitude, position, or value of something)[1]

TO ANALYZE: to examine methodically by separating into parts and studying their interrelations[2]

Now, for the nouns:

ESTIMATE: a tentative evaluation or rough calculation, as of worth, quantity, or size; a statement of the approximate cost of work to be done, such as a building project or car repairs[3]

ANALYSIS: the separation of an intellectual or material whole into its constituent parts for individual study; the study of such constituent parts and their interrelationships in making up a whole.[4]

---

1. http://dictionary.reference.com/search?q=estimate

2. http://dictionary.reference.com/search?q=analyze

3. http://dictionary.reference.com/search?q=estimate

4. http://dictionary.reference.com/search?q=analysis

It is clear that estimation is about quantification of cost, whereas analysis is about the decomposition and integration of a set of parts. Analysis adds value by creating knowledge about what components we need to create and how they integrate together to synthesize a whole solution. Estimation merely quantifies the cost. I believe that in our industry, people are lousy at estimating. And that this may indeed be true of all knowledge work. Estimating knowledge work is hard. It's even harder without proper analysis. However, reliable and repeatable analysis techniques—although they exhibit variation, and are seldom used perfectly—significantly contribute to our understanding and planning of software development and project delivery, do exist, and should be used. Analysis reduces variation and makes plans more accurate. Estimation does neither of these things. Agile estimating,[1] on the other hand, uses historical productivity data, and it uses the output of analysis to provide a reliable estimate with a buffer for variation. The result is a plan that is more accurate, and the time spent on it was value adding, knowledge creating, and decomposition and integration, rather than inaccurate, non-value added cost quantification.

1. http://www.agilemanagement.net/index.php/blog/Agile_Estimating

Once in an Agile Lifetime

[With apologies to fellow Scot, David Byrne, and Talking Heads]
Sunday, February 19, 2006

<!—StartFragment—>And you may find yourself in a two-bit startup
And you may find yourself in another part of your career
And you may find yourself in control of a large development team
And you may find yourself in a beautiful team, with a beautiful project
And you may ask yourself—well . . . how did I get here?

Letting the days go by/letting the defects hold me down
Letting the days go by/defects choking up the schedule
Into the blue again/after the budget's gone
Once in a lifetime/defects chocking up the schedule.

And you may ask yourself
How do I work this?
And you may ask yourself
Where is that beautiful process?
And you may tell yourself
This is not my beautiful team!
And you may tell yourself
This is not my beautiful project!

Letting the days go by/letting the defects hold me down
Letting the days go by/defects choking up the schedule
Into the blue again/after the budget's gone
Once in a lifetime/defects chocking up the schedule.

Same as it ever was . . . same as it ever was . . . same as it ever was . . .
Same as it ever was . . . same as it ever was . . . same as it ever was . . .
Same as it ever was . . . same as it ever was . . .

Bugs dissolving . . . and code refactoring
There is value at the bottom of the iteration

Carry the change request through the end of the iteration
Remove the bugs at the bottom of the iteration!

Letting the days go by/letting the defects hold me down
Letting the days go by/defects choking up the schedule
Into the blue again/in the silent standup
Under the build report/there defects choking up the schedule.

Letting the days go by/letting the defects hold me down
Letting the days go by/defects choking up the schedule
Into the blue again/after the budget's gone
Once in a lifetime/defects choking up the schedule.

And you may ask yourself
What is that beautiful project?
And you may ask yourself
Where did that schedule go?
And you may ask yourself
Am I right? . . . or am I wrong?
And you may tell yourself
My god! . . . what have I done?

Letting the days go by/letting the defects hold me down
Letting the days go by/defects choking up the schedule
Into the blue again/in the silent standup
Under the build report/there defects choking up the schedule.

Letting the days go by/letting the defects hold me down
Letting the days go by/defects choking up the schedule
Into the blue again/after the budget's gone
Once in a lifetime/defects choking up the schedule.

Same as it ever was . . .same as it ever was . . .same as it ever was . . .
Same as it ever was . . .same as it ever was . . .same as it ever was . . .
Same as it ever was . . .same as it ever was . . .

CHAPTER 12

Defining Agile Leadership

I'VE CHOSEN TO INCLUDE THIS CHAPTER TO PROVIDE SOME HISTORY ON AGILE PROJECT MANAGEMENT AND CERTIFICATION, AS WELL AS TO SHOW AN EXAMPLE OF AN ATTEMPT TO START A MOVEMENT. Early in 2004, I received a phone call from Jim Highsmith. He told me that he wanted to pull a group of people together to create an equivalent of the Agile Manifesto, but focused on project management. He wanted this group to be diverse and to include many thought leaders and authors from outside the software world. He was impressed with my book, *Agile Management for Software Engineering* (Prentice Hall, 2003), and wanted me to be part of it.

What we created in 2005 is now known as the Agile Leadership Network, and it has a few local chapters, most notably in San Francisco. It still occasionally holds a conference, the most recent of which was in Houston, Texas. However, in my opinion, the APLN (as it was known) failed on its basic objectives—failed to gain much of a following or to gain traction in the wider Agile community, and did not catalyze a wider movement in modern project management. We can learn a lot from this failure, and I did.

When forming the Lean Software & Systems Consortium, now the Lean Systems Society, my experience with the APLN was invaluable. In growing the Kanban movement and its related institutions, Lean Kanban University and Limited WIP Society, I found that my experience with APLN and its events and local chapters was invaluable. However, a full analysis of what went well with the APLN, and how, in my opinion, it failed, together with my thoughts on how to create a successful movement, needs to wait for another book.

I felt it was right and fair that someone should capture the intellectual output from the founders of the APLN and publish it for posterity. Very few people in the Agile community know about the APLN, and fewer are aware that its values and principles were captured in a document, pompously titled "The Declaration of Interdependence."

Test-First Meeting Facilitation

Wednesday, February 2, 2005

• •

Asking people to define the veto hurdle focused
each person on what was most important to them.

• •

THESE PAST THREE DAYS, I'VE BEEN HOSTING A GROUP OF AUTHORS AND luminaries from the Agile project management community. You'll be hearing a lot more about this meeting all across the Internet over the next few days and weeks. I'll be saying more about what we discussed and the outcome of the meeting next week (or later this week in my Yahoo! group). However, I thought I'd pause, tonight, to mention not what we discussed, but how we discussed it—test-first meeting facilitation.

The idea really came (inadvertently) from Todd Little, a board member of the Agile Alliance, who kept asking, "How will we know when we're finished discussing [this important topic]?" The answer was to devise a test for completeness of the topic. So, as we started each agenda item, we identified the test (or checklist) against which the output must be measured. Perhaps an hour or two later, we'd be at a point of consensus, then we'd say, "Now, let's check this against our test to see if we're done." If it passed the test, and everyone in the room had a thumb up, it was recorded in the minutes as agreed. If it failed the test, we kept discussing the outstanding points until we had a consensus agreement that passed the test.

The test represented the set of minimum criteria with which each participant in the session could live, that is, it had to pass their test, or they would veto the whole thing. Asking people to define the veto hurdle focused each person on what was most important to him or her. This made the accumulated test sufficient without being bloated. When an agenda item passed the test, a consensus was reached (by default) because each participant's veto threshold had passed.

A simple but powerful mechanism—and one that I'll be using again.

Declaration of Interdependence

Sunday, February 6, 2005

● ●

Melt down that iron triangle and learn to facilitate
the flow of value!

● ●

HAS EXISTING PROJECT MANAGEMENT GUIDANCE BEEN ANNOYING YOU WITH ITS failure to help when you're faced with uncertainty and change? Have you been doing your best, but it just isn't working? Are you looking for something new to help make you successful? Then maybe it's time for you to subordinate your *Guide to the Project Management Body of Knowledge* (PMBOK) and embrace uncertainty! Melt down that iron triangle and learn to facilitate the flow of value! The Declaration of Interdependence is here! It's new! It's modern! It's progressive! It's a movement! Show your allegiance—sign the declaration! Let's lead the future of project management together.

❒ We increase return on investment by making continuous flow of value our focus.

❒ We deliver reliable results by engaging customers in frequent interactions and shared ownership.

❒ We expect uncertainty and manage for it through iterations, anticipation, and adaptation.

❒ We unleash creativity and innovation by recognizing that individuals are the ultimate source of value, and creating an environment where they can make a difference.

❒ We boost performance through group accountability for results and shared responsibility for team effectiveness.

❒ We improve effectiveness and reliability through situationally specific strategies, processes, and practices.

©2005 David Anderson, Sanjiv Augustine, Christopher Avery, Alistair Cockburn, Mike Cohn, Doug DeCarlo, Donna Fitzgerald, Jim Highsmith, Ole Jepsen, Lowell Lindstrom, Todd Little, Kent McDonald, Pollyanna Pixton, Preston Smith, and Robert Wysocki.

(This is the fruit of almost a year of discussion, which led to a two-day meeting last week in Redmond, Washington. Getting 15 thought leaders in management and agility to agree was actually easier than you might think. We all agreed on the paradigm. The hard work was the wording. By the good fortune of alphabetical order, it seems I get to be the first named signatory. Hmmm. That's nice.)

Thoughts on the DOI

Monday, February 14, 2005 through Saturday, February 19, 2005

●●●●●●●●●●●●●●●●●●●●●●●●●●●●●●●●●●●

"Partner" means to align the supply chains' interests and focus on the end consumer. This is how Japanese *keiretsu* work. It creates a mechanism to treat the whole value chain as a single system, and optimize for the system, not the parts.

●●●●●●●●●●●●●●●●●●●●●●●●●●●●●●●●●●●

THIS REPRESENTS MY OWN VIEWS, AND NOT ANY OFFICIAL VIEW FROM MY co-signatories.

1. Flow

The first statement in the declaration says:

❏ We increase return on investment by making continuous flow of value our focus.

What this means for me is, first, we need to treat projects as a flow problem. This is concomitant with Peter Drucker's idea of a value chain. In a (software) project, the value chain creates consumer value by transforming ideas into working knowledge through a series of transformative steps. This idea was fundamental to the thesis of my book, *Agile Management for Software Engineering* (Prentice Hall, 2003), and at the time I wrote it, I spent rather a lot of words laboring the point in the early chapters. It seems like a much more accepted principle nowadays, but back then I felt I had to do a lot of justifying to get this principle built into the management paradigm.

So, that takes care of the latter half of the sentence, but what about the first half?

Well, the argument goes, until you embrace the idea of flow through a value chain, you cannot understand where to focus management attention and investment dollars in order to maximize the investors' return for those dollars. Hence, adopting a flow paradigm is fundamental to devising optimal investment strategies.

A flow model also reveals the primary role for the agile project manager—issue-log management and resolution. By focusing on what is obstructing flow—issues arising—the project manager keeps things moving. The Scrum community will identify with this, as it is the primary role of the Scrummaster. Ken Schwaber teaches quite clearly that he sees, as the primary mechanism for realizing results, managing the issue log, surfacing issues at daily scrums, and running them down before they impact "burn down."

Astute readers who are fans of the Theory of Constraints will recognize that this first bullet also encompasses and embraces the underlying principles of TOC and its Five Focusing Steps. Once you have a flow model, you can look for bottlenecks in the flow.

2. Customers

❐ We **deliver reliable results** by engaging customers in frequent interactions and shared ownership.

We wanted to capture the idea of a close partnership with customers. The term partnership tends to get misused. In the same way that executives want you to "delight customers," often, they also want you to "partner" with them, without fully understanding what that means. "Partner" means to align the supply chains' interests and focus on the end consumer. This is how Japanese *keiretsu* work. It creates a mechanism to treat the whole value chain as a single system, and optimize for the system, not the parts.

We also wanted to capture the quality assurance that comes from frequent customer contact and feedback.

Together, these two aspects of putting the customer's skin in the game through (inter-)active partnership and frequent contact gives us a mechanism to deliver reliable results. The word "reliable" embraces concepts such as repeatable, dependable, and trustworthy, without the historical baggage that the word "repeatable" carries in process circles.

3. Uncertainty

❐ We expect uncertainty and manage for it through iterations, anticipation, and adaptation.

The key term here is "uncertainty," and its scope. We had to settle on one word to capture the essence of variation, change, unknowns, and chaos. We settled on uncertainty as the single word that best captures the essence of the problem. Uncertainty, in this context, means from variation to chaos, as defined by De Meyer et al, in "Managing Project Uncertainty: From Variation to Chaos" (*Sloan Management Review,* Winter 2002), which is illustrated in Figure 12.1.

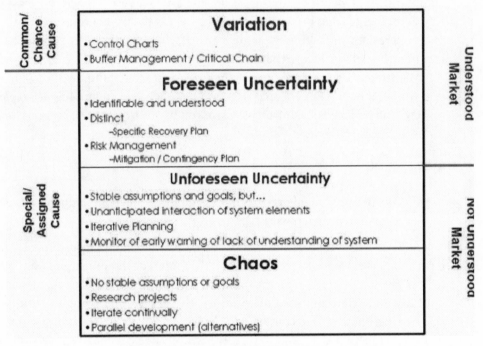

Figure 12.1 The continuum from variation to chaos

By embracing uncertainty in the new paradigm, we're clearly making a split from traditional project-management theory. There is no concept of uncertainty in Critical Path. In the PMBOK, uncertainty is handled through the notion of risk. Variation, in the latest, 2004 edition, is handled through positive and negative risk

The DOI embraces uncertainty. It's a reality of the universe we live in. It's fundamental. No model of project management can deny it or go without it.

We deal with uncertainty by iterating often, and by providing control and feedback points to make corrections based on the effects of uncertainty. Also, we anticipate uncertainty. It is the inclusion of "anticipate" in this statement that is important and differentiating from most existing Agile literature. Agile methods tend to be reactionary to unfolding events. The authors of the DOI are encouraging you to be more anticipatory. Anticipation should enable you to react more swiftly and to show greater agility.

Finally, we learn from our feedback and adapt. This can mean adjusting future anticipation, or simply reacting to current events with adaptive planning. In the new paradigm, uncertainty is a facet of planning and scheduling. Planning can be anticipatory or reactive, or both.

4. Innovation

❑ .We unleash creativity and innovation by recognizing that individuals are the ultimate source of value, and creating an environment where they can make a difference.

We wanted to capture something in the DOI that expressed the notion that we view individuals as assets in a business, rather than as faceless resources that are somehow magically interchangeable and fungible in a three-sided iron triangle tradeoff. Saying that people are the ultimate source of value is a recognition that modern work is knowledge work. Knowledge is value. People create that knowledge. The creation of new knowledge is an act of creativity that generates innovation. This happens more successfully when you create an environment where people are viewed as assets and are encouraged to take risks and make a difference. This isn't really a project-management practice, rather, it's a good modern-management practice

5. Team

❑ We boost performance through group accountability for results and shared responsibility for team effectiveness.

Taylorism in the twentieth century was all about the cost-accounting notion that you optimize the whole by focusing on efficiency at the individual level. With the DOI, we are calling this out as wrong. We're taking a Peter Drucker/ Michael Porter–style view that business is done in a value chain. Optimal performance is reached when everyone in the value chain partners together and has a vested interest in the success of the whole chain. It is this notion that led us to the statement "shared responsibility for team effectiveness."

"Group accountability" captures the notion of "personal safety." Personal safety is the flip side of courage. Why would you need to be courageous? Because you do not live in an environment of personal safety! Hence, group accountability says, "We are all in this together. Together we succeed or together we accept responsibility for our failure." By creating an environment of personal safety, we believe that individuals are motivated to do the right thing for the optimal performance of the whole system—and not to make a local decision to protect themselves, which is sub-optimal to the global performance.

6. Situationally Specific

❐ We improve effectiveness and reliability through situationally specific strategies, processes, and practices.

For me, this was the most difficult to accept of all six statements. I had wanted something stronger—recognition that the most effective solution couples the engineering discipline with the project-management discipline and allows them to feed off each other intelligently. I've learned this from FDD. The project-management aspects of FDD are different because the Coad Method of software engineering enables a different way to think about project management. Project management in FDD isn't some generic task-centric formula. It's based on the software architecture and domain and requirements analysis method. The reporting isn't standard earned value; rather, it's based on the production of features. I believe that when project management and engineering disciplines are combined, a more effective process is delivered. I didn't managed to get that included in the Declaration, so I settled for this message about situationally specific opportunities, which gets us most of the way there.

So, for me, the DOI is embracing the idea that it is okay to argue for different approaches, and to customize project-management techniques, and to couple those methods with knowledge of how the engineering technique works. It's okay—if you can show that it leads to a more economically effective solution.

CMMI and the Declaration
of Interdependence

Thursday, July 7, 2005

THE DOI WAS BORN OUT OF OBSERVING THAT PROJECT-MANAGEMENT PRAC-tices, as taught in so many places, are leading to undesirable behavior and unfavorable results. The DOI seeks to reset the mind set (or paradigm) of how people think about projects so that they adopt the correct behavior that leads to good results. Back in February, I was involved in the creation of a rallying cry for the adoption of Agile techniques in project management. We called this the Declaration of Interdependence—six statements that we felt differentiated the Agile approach from traditional teaching of project management. Shortly after publication, I was challenged (via my Yahoo! group) to state how the DOI differs from existing project-management frameworks, such as the CMMI IPPD from the Software Engineering Institute. My initial reaction was to state that because the CMMI is a framework that doesn't prescribe specific practices, it was likely to be compatible, and that, despite frameworks like the CMMI, it was still possible that many project managers were practicing techniques that were damaging. I promised to do an analysis, and to post it here when it was done.

After nine months of being deeply immersed in the CMMI while producing MSF for CMMI® Process Improvement, including two visits to the SEI at Carnegie Mellon University in Pittsburgh, Pennsylvania, I'm finally ready to make good on the promise—a gap analysis of the DOI versus the CMMI.

[DOI] We increase return on investment by making continuous flow of value our focus.

[CMMI] The CMMI is pretty agnostic on this. It's a "so what?" It doesn't break the CMMI, nor does the CMMI ask us to do anything differently. Although the CMMI is written in a project-centric fashion, it does not rule out small increments of delivery.

#1 is compatible!

[DOI] We deliver reliable results by engaging customers in frequent interactions and shared ownership.

[CMMI] The CMMI actively encourages stakeholder involvement, and it has explicit activities for monitoring it.

#2 is compatible!

[DOI] We expect uncertainty and manage for it through iterations, anticipation, and adaptation.

[CMMI] The CMMI is founded on the principles of W. Edwards Deming, his Theory of Profound Knowledge, the theory of variation, and the concepts of special- and common-cause variation. Although some explicit text talks about "plan the plan, plan the work, work the plan" and has very "conformance to plan" language in some of the practice guidance, there is nothing about uncertainty, variation, adaptive planning, iterations, and anticipation that is incompatible with CMMI. In fact the new MSF for CMMI® Process Improvement does these things explicitly.

#3 is compatible!

[DOI] We unleash creativity and innovation by recognizing that individuals are the ultimate source of value, and creating an environment where they can make a difference.

[CMMI] The CMMI is very big on the idea of creating an organizational environment for success (where success is defined as achieving continuous improvement—in a quality-assurance sense). There are explicit process areas around training and the organizational environment. There is nothing in the CMMI that is incompatible with the DOI's embrace of innovation and creativity and its underlying principle that people are assets rather than fungible cost centers.

#4 is compatible!

[DOI] We boost performance through group accountability for results and shared responsibility for team effectiveness.

[CMMI] Although many formal organizations that follow the CMMI use RACI designations, and assign individual accountability, the CMMI is really agnostic about this. What it asks for is that the agreement is written in a Team Charter. Again, with MSF for CMMI® Process Improvement, I have designed it to exhibit group accountability and shared responsibility with the "team of peers" concept carried over from earlier versions of MSF. This is perfectly compatible with CMMI.

#5 is compatible!

[DOI] We improve effectiveness and reliability through situationally specific strategies, processes, and practices.

[CMMI] The CMMI explicitly expects situationally specific practices and processes, and it encourages them with explicit activities aimed at tailoring processes and practices for specific projects.

#6 is compatible!

Asking the question the other way, "Is there anything in the CMMI that is incompatible with the DOI?" is also interesting. There are 25 process areas in the CMMI. However, my take on it is, "No! There is nothing within the process areas or the generic and specific goals of the CMMI model that are incompatible with the DOI." The problems arise in how people interpret the specific practices. The DOI was born out of observing that project-management practices, as taught in so many places, are leading to undesirable behavior and unfavorable results. The DOI seeks to reset the mindset (or paradigm) of how people think about projects so that they adopt the correct behavior that leads to good results.

It's perfectly possible to be an agile project manager and run a CMMI-compliant process.

Kanban and the Declaration
of Interdependence

Tuesday, December 21, 2010

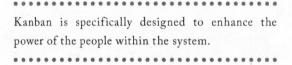

Kanban is specifically designed to enhance the power of the people within the system.

In 2004, I was involved with a group of recognized professionals in the field of project management who came together to define what agile project management ought to mean. The outcome of those meetings was a statement of values, published in February 2005, named the Declaration of Interdependence. Although I'm not in love with the name (I think it's rather pompous), I find that the content of the Declaration has stood the test of time over the five years since we wrote it. The declaration sought to define a value system by which modern, twenty-first–century project managers should work, but its secondary purpose—to galvanize a community around general application of agile project management—failed to materialize. Five years on, it is worth reflecting on my contribution to the DOI and how it aligns with the Kanban work that I am most closely associated with now.

❏ We increase return on investment by making continuous flow of value our focus.

As many of my readers are aware, kanban systems limit work-in-progress, and signal new work to begin only when there is capacity to process it. This "pull" mechanism is used to improve the flow of work. So, kanban systems are focused on flow. However, kanban systems for software development have evolved well beyond a typical manufacturing implementation, as Don Reinertsen pointed out to me during a visit to my office in 2007. In the Kanban Method, we use classes of service linked to (opportunity) cost of delay and explicit visualization of handling policies so as to improve the return on investment made to support operating the software-development activity. By visualizing the workflow, limiting WIP, managing flow, measuring lead times, and optimizing both risk and value delivery with classes of service based on the economics of delay—Kanban explicitly delivers on the first statement of the DOI. It is no surprise that this first of the six statements is so heavily associated with my contribution.

> ❐ We deliver reliable results by engaging customers in frequent interactions and shared ownership.

Kanban engages customers through visualization, through interaction and escalation when items are blocked, through the regular cadence and collaboration of input-queue replenishment meetings, and through the regular cadence and the planning of each delivery. Kanban asks customers to take shared ownership of the system and its effectiveness, and to throttle their demand to the rate at which the system can deliver.

> ❐ We expect uncertainty and manage for it through iterations, anticipation, and adaptation.

Kanban embraces uncertainty, and it manages for it by providing of classes of service, such as "Expedite." This demonstrates anticipation of demand. It also manages for uncertainty by using quantitative measurement, such as statistical process control and definition of target lead times based on statistical observation of capability. This, again, shows how the system can anticipate an outcome and manage for variability and uncertainty. Kanban is also designed to encourage adaptation by using the WIP limit and the social interaction of standup meetings and operations reviews to reflect on the system's performance, observed capability and effectiveness. It also allows for process improvements to be suggested, based on a scientific approach that uses models of process flow, variation, uncertainty, and complexity to facilitate implementation of successful adaptations. The Kanban Method is specifically designed to anticipate and adapt, as it is designed based on concepts of both Systems Thinking and Complex Adaptive Systems.

However, depending on how you define "iterations," Kanban implementations often do not use them! Time-boxed increments (often referred to as "iterations" by agile practitioners) are replaced with cadence. The core activities of accepting new work and delivering finished work are still usually performed regularly, but each activity has its own cadence. For example, input queue replenishment might happen once per week—a weekly cadence—and delivery might happen every second week—a bi-weekly cadence. Cadence is a more sophisticated tool than time-boxed iterations.

"Iteration," meaning "to rework or refine," is supported by Kanban. However, it is still relatively unusual to see such implementations. If, for example, a team follows an explicitly iterative method, such as Barry Boehm's Spiral Model, it would be possible to wrap a kanban system over that and to limit WIP at each step on each loop of the spiral. There have even been instances of teams visualizing such a process with a board that resembles a dartboard or archery target rather than the familiar columns and rows typically associated with Kanban.

❑ We unleash creativity and innovation by recognizing that individuals are the ultimate source of value, and creating an environment where they can make a difference.

Kanban is specifically designed to enhance the power of the people within the system. By limiting WIP, kanban systems create slack, which allows creativity and innovation the room to sprout and flourish. Additionally, the act of making process policies explicit provides protection for those working within the system, limiting abuses and attempts to exploit individuals and work around limitations. The explicit policies, often visualized on a board, enable individual team members to make high-quality decisions about the economics, risks, and intangible expectations of all stakeholders. As a result, Kanban enables individuals to be creative about their processes, to innovate and adapt to improve customer service, and to deliver value. It empowers them to make their own decisions and to optimize the economic outcome for the benefit of customers, business owners, value-stream partners, and other team members.

❑ We boost performance through group accountability for results and shared responsibility for team effectiveness.

Kanban uses a monthly operations review meeting that involves all team members across an organization, plus senior management, and up- and downstream stakeholders. At operations review, quantitative, objective, and statistical data on the performance and capability of the system is reviewed openly by all involved; design changes to the system are proposed and assigned to managers for implementation. Through this operations review process, Kanban ensures that performance is boosted continuously because the wider group of stakeholders takes shared responsibility for the effectiveness of the system and for the team of people operating it.

❑ We improve effectiveness and reliability through situationally specific strategies, processes, and practices.

The first emergent property of Kanban is that each process is uniquely tailored to its context. For each system, the value-creation network is different, the risk profile is different, the team and its skills and capabilities is different, the nature of demand is different, and the cost of delay in work items is different. Every context deserves a uniquely tailored process in order to optimize the economic outcome for all stakeholders. The foundational Principles of the Kanban Method are based on the premise that process definition starts with whatever is happening now, and it evolves incrementally through a process of small changes—each justified economically, and based on a scientific approach to improving performance. Situationally specific strategies, process, and practices are core to the Kanban Method.

Conclusions

Not only is it easy to see how the Kanban Method relates to the Declaration of Interdependence, and to see how my emerging "agile" work at the time was aligned with it, we could go further and say that Kanban is, in fact, a full implementation of the Declaration of Interdependence. It is a prescription for how twenty-first–century project managers ought to be working. The irony of this might be that Kanban encourages a service-delivery approach to work, rather than a project-centric one.

Meanwhile, for me, personally, I look back on the Declaration of Interdependence as a piece of work that I and the other authors can rightly be proud of. I am also happy that this review of the Kanban Method and how my work has evolved in the intervening years shows a high level of consistency, and that Kanban demonstrates that it lives the values defined in the Declaration.

I've been guilty of not publicizing the Declaration of Interdependence enough. I'm not the only author who fails to make enough use out of it. It rightly is something to be proud of, and it deserves a higher profile within the Agile and Lean/Kanban communities. What a pity about its name!

The DOI, Made to Slip?

By Mike Griffiths, Leading Answers Blog, December 13, 2007
(used with kind permission)

• •

How do we get people to act on our ideas? We tell
stories.

• •

Nearly three years on, why is the Declaration of Interdependence (DOI) still not widely known or referenced?

Chances are that most readers are not familiar with the DOI, yet it is a great piece of work. The DOI lists principles that, like the Agile Manifesto's principles, help point the way for teams working on agile projects. It was created by the founders of the Agile Project Leadership Network (APLN) to guide agile project management and rally support for an uprising of new project-management thinking.

Other than believing that some of the wording was a little too clever for its own good, or for general consumption, I did not fully understand why it had been avoided. Then I read *Made to Stick,* by Chip and Dan Heath, and I realized that the DOI has the stickiness and appeal of a greased electric eel. *Made to Stick* is a great book that practices what it preaches and conveys how to make messages effective in an enjoyable and memorable way. The mental model the authors use for explaining why some stories are unforgettable is the hooks-and-loops idea of Velcro. Imagine that our brains have millions of tiny loops to store information; ideas need hooks to latch onto these loops, otherwise they will slide by or be dropped and forgotten.

Some stories, such as urban legends and proverbs, have many hooks, and so they stick in people's minds very well. They are said to be "sticky ideas,"; they endure, they spread, and they are effective. Consider the following classic from the book:

> A friend of a friend of ours is a frequent business traveler. Let's call him
> Dave. Dave was recently in Atlantic City for an important meeting with
> clients. Afterward, he had some time to kill before his flight, so he went
> to a local bar for a drink. He'd just finished one drink when an attrac-
> tive woman approached and asked if she could buy him another. He was

surprised but flattered. Sure, he said. The woman walked to the bar and brought back two more drinks—one for her and one for him. He thanked her and took a sip. And that was the last thing he remembered.

Rather, that was the last thing he remembered until he woke up, disoriented, lying in a hotel bathtub, his body submerged in ice. He looked around frantically, trying to figure out where he was and how he got there. Then he spotted the note: Don't move. Call 911.

A cell phone rested on a small table beside the bathtub. He picked it up and called 911, his fingers numb and clumsy from the ice. The operator seemed oddly familiar with his situation. She said, "Sir, I want you to reach behind you, slowly and carefully. Is there a tube protruding from your lower back?"

Anxious, he felt around behind him. Sure enough, there was a tube. The operator said, "Sir, don't panic, but one of your kidneys has been harvested. There's a ring of organ thieves operating in this city, and they got to you. Paramedics are on their way. Don't move until they arrive."

Why do stories like this spread like wildfire and lodge in our brains when real, useful information never gets in, or slides right by? Well, because they are sticky; they contain the hooks our brains are hard-wired to latch onto. These hooks are:

- ❏ SIMPLICITY

- ❏ UNEXPECTEDNESS

- ❏ CREDIBILITY

- ❏ EMOTIONS

- ❏ STORIES

We will look at these hooks in more detail, and the DOI in a little while; but first, another concept we need to understand is the "Curse of Knowledge."

Stripped Bear of Hooks by "The Curse of Knowledge"

The Curse of Knowledge is name given to the problem that when people are "in the know" about a subject, information and details seem obvious, and there is no need to restate them. However, people who are not in the know need these details to grasp and remember the idea.

This, I believe, is the big problem with the DOI. It was created by a group of people who "got-it" and it was written in their high-level shorthand, stripped of supporting detail so as to keep ideas as compact as possible. It makes perfect, logical sense that they should create the recommendations like this; they are the most succinct way of conveying important, but complex ideas. However, this insider, shorthand code lacks any of the hooks that would make them easily understood and memorable.

Can We Fix It? Yes, We Can!

It is easy to pick holes at something, but it is more challenging (and useful) to try to improve it. So, how do we make the DOI stickier? First, we should understand the definitions of Sticky Hooks, and then assess the DOI against them.

Made to Stick Principles

From the *Made to Stick* authors, here are some further explanations of the Sticky principles:

PRINCIPLE 1: SIMPLICITY

How do we find the essential core of our ideas? A successful defense lawyer says, "If you argue ten points, even if each is a good point, when they get back to the jury room they won't remember any." To strip an idea down to its core, we must be masters of exclusion. We must relentlessly prioritize. Saying something short is not the mission—sound bites are not the ideal. Proverbs are the ideal. We must create ideas that are both simple and profound. Find the core.

PRINCIPLE 2: UNEXPECTEDNESS

How do we get our audience to pay attention to our ideas, and how do we maintain their interest when we need time to get the ideas across? We need to violate people's expectations. We need to be counterintuitive. We can use surprise—an emotion whose function is to increase alertness and cause focus— to grab people's attention. But surprise doesn't last. For our idea to endure, we must generate interest and curiosity. How do you keep students engaged during the forty-eighth history class of the year? We can engage people's curiosity over a long period of time by systematically "opening gaps" in their knowledge—and then filling those gaps. Motivate the people to pay attention by seizing the power of surprise and highlight a knowledge gap.

PRINCIPLE 3: CONCRETENESS

How do we make our ideas clear? We must explain our ideas in terms of human actions, in terms of sensory information. This is where so much business communication goes awry. Mission statements, synergies, strategies, visions—they are often ambiguous to the point of being meaningless. Naturally sticky ideas are full of concrete images—ice-filled bathtubs— because our brains are wired to remember concrete data. In proverbs, abstract truths are often encoded in concrete language: "A bird in hand is worth two in the bush." Speaking concretely is the only way to ensure that our idea will mean the same thing to everyone in our audience.

PRINCIPLE 4: CREDIBILITY

How do we make people believe our ideas? When the former surgeon general C. Everett Koop talks about a public-health issue, most people accept his ideas without skepticism. But in most day-to-day situations we don't enjoy this authority. Sticky ideas have to carry their own credentials. We need ways to help people test our ideas for themselves. When we're trying to build a case for something, most of us instinctively grasp for hard numbers. But in many cases this is exactly the wrong approach. In the sole U.S. presidential debate in 1980 between Ronald Reagan and Jimmy Carter, Reagan could have cited innumerable statistics demonstrating the sluggishness of the economy. Instead, he asked a simple question that allowed voters to test for themselves: "Before you vote, ask yourself if you are better off today than you were four years ago." Boost credibility with vivid details.

PRINCIPLE 5: EMOTIONS

How do we get people to care about our ideas? We make them feel something. Research shows that people are more likely to make a charitable gift to a single needy individual than to an entire impoverished region. We are wired to feel things for people, not for abstractions, so appeal to self interest.

PRINCIPLE 6: STORIES

How do we get people to act on our ideas? We tell stories. Firefighters naturally swap stories after every fire, and by doing so they multiply their experience; after years of hearing stories, they have a richer, more complete mental catalog of critical situations they might confront during a fire and the appropriate responses to those situations. Research shows that mentally rehearsing a situation helps us perform better when we encounter that situation in the physical environment. Similarly, hearing stories acts as a kind of mental flight simulator, preparing us to respond more quickly and effectively.

The Declaration Of Interdependence (DOI)

So, how does the DOI measure up?

(For brevity, I discuss the collection of DOI statements under each heading, but to be effective, each statement should hold up to all the tests.)

SIMPLICITY

The statements are short, which is good. However, they are not very simple. For instance, we have to read the first principle right to the end to find out we are talking about "focus" and then reread to work out what it is we are supposed to focus on. The use of jargon (return on investment), insider terms (flow of value, iterations) and questionable English (situationally) further complicates the messages.

UNEXPECTEDNESS

The statements do not have a lot of unexpected content to catch our attention. "We expect uncertainty" is probably the best candidate, but it's hardly "Man bites dog."

CONCRETENESS

The statements are largely abstract, and they are stripped of practical implementation details. This keeps them short, but it makes them harder for people to relate to real-world situations.

CREDIBILITY

Respected industry experts created the DOI. Their signatures, and those of others supporting the work, add credence to the recommendations. However, they lack vivid details that show how they would work, or a Sinatra test (well, if it works here, it would work anywhere).

EMOTIONS

The statements start with the word "We." That makes them a little more human and friendly, but they otherwise lack personal connection or emotion.

STORIES

The DOI lacks stories, so, it does not allow us to simulate the circumstances in our heads.

An Expanded DOI

1. We increase return on investment by making continuous flow of value our focus.

A stickier version might be:

1. Amaze your users, keep giving them what they ask for!

Concentrate on developing features the business asks for. This is how we can get the best business benefits and support for the process. When your projects consistently deliver business results, they are hard to ignore, cancel, or deny requests from.

2. We deliver reliable results by engaging customers in frequent interactions and shared ownership.

A stickier version might be:

2. When planning interaction with the business, try to be more like the good neighbor whom you see frequently and can easily call on, rather than the intrusive relative who moves in for a while and then disappears for a year. We want business interaction that is regular and engaging, not a huge up-front requirements-gathering phase followed by nothing until delivery. Frequently show how the system is evolving and make it clear that the business drives the design by listening and acting on the feedback.

3. We expect uncertainty and manage for it through iterations, anticipation, and adaptation.

A stickier version might be:

3. Because software functionality is hard to describe, and both technology and business needs change quickly, software projects typically have lots of unanticipated changes. Rather than trying to create and follow a rigid plan that is likely to break, it is better to plan and develop in short chunks (iterations) and adapt to changing requirements.

Story: The first .NET project I managed involved building an online drugstore for a Canadian pharmacy. Selling drugs to the US was a contentious business, with regulations and business rules changing very frequently (weekly). The client was trying to overtake rival online pharmacies, so there was an arms-race of special offers, loyalty schemes, and promotions to win customers that changed frequently and, since .NET was still very new, it was changing also.

All this change and uncertainty meant that detailed plans and long release cycles just did not work. However, by adopting an agile approach—with daily iterations and weekly releases of the live site—the client was able to respond to changing business circumstances and overtake less agile competitors, gaining considerable market share.

4.   We unleash creativity and innovation by recognizing that individuals are the ultimate source of value, and creating an environment where they can make a difference.

A stickier version might be:

4.   We manage property and lead people; if you try to manage people, they feel like property.

Projects are completed by living, breathing people, not tools or processes. To get the best out of your team, you must treat them as individuals, provide for their needs, and support them in the job. Paying people a wage might guarantee that they show up, but how they contribute once they are there is governed by a wide variety of factors. If you want the best results, provide the best environment.

5.   We boost performance through group accountability for results and shared responsibility for team effectiveness.

A stickier version might be:

5.   Everyone needs to share responsibility for making the project, and the team and as a whole, successful.

6.   We improve effectiveness and reliability through situationally specific strategies, processes and practices.

A stickier version might be:

6.   Real projects are complex and messy; rarely do all the ideal conditions of agile development present themselves. Instead, we have to interpret the situation and make the best use of the techniques, people, and tools available to us. There is no single cookbook for how to run successful projects; instead, we need to adjust for the best fit with the project ingredients and environment we are presented with.

The Expanded DOI Stickiness Test

So how do these extensions measure up?

SIMPLICITY

The statements are not short anymore, but then, neither is the bathtub story. They try to layer the explanation of concepts through a series of simple statements. I try to start with the main point (don't "bury the lead") and then add supporting detail.

UNEXPECTEDNESS

There is not a lot of unexpectedness in the expanded text either. The "property and people" quote might be new to some people.

CONCRETENESS

I think increasing the number of examples and stories helps make them more concrete. We can imagine all the changes needed on the drugstore project, and the friendly neighbor versus the invasive houseguest metaphor is something everyone can relate to.

CREDIBILITY

Accounts of real-world examples where agile worked in spite of adversity (drugstore project) add credibility by way of the Sinatra principle (" . . .if I can make it there, I'll make it anywhere . . ."), and by appealing to common sense.

EMOTIONS

By introducing self interest ("When your team consistently delivers . . .," "To get the best out of your team . . .") readers will relate the statements more to how they can be of use to them. This "useful-to-me" feeling should help boost retention.

STORIES

The online-pharmacy story helps illustrate some of the principles, but other stories would be good too.

Summary

So, while originally the DOI was "Made to Slip," perhaps it can be "Refactored to Stick" and get the attention it deserves. Figure 12.2 illustrates the differences between the two.

Mike does a good job of identifying why the DOI didn't catalyze a movement. It wasn't the only factor in why the APLN failed to generate the change in the industry that the authors hoped it would. Much more could be written about the APLN, the challenges of synthesizing intellectual property from 12 or more authors into a single standard, and the failure to agree on a mechanism for spreading education and providing professional recognition. APLN lacked a viable business model—it lacked both emotional drivers to galvanize a movement and it lacked economic drivers.

Reading Mike's analysis and examining the *Made to Stick* framework developed by Chip and Dan Heath, it is easy to see why Kanban has done so much better and why the Kanban community and movement has thrived over the past five years: Kanban is simple, it is unexpected, it is concrete, it has credibility, it connects with people emotionally through the visual and tactile nature of boards and through the recognition that resistance to change is emotional, and through a respect for the tribal nature of people, and finally, Kanban has stories –lots of stories. I've actively encouraged storytelling and have created forums such as the Lean Kanban series of conferences for people to tell their stories.

My failed experience with the APLN strengthened me and gave me understanding that has enabled Kanban to thrive and grow. Meanwhile, the DOI stands as a solid foundation as a set of principles for modern management.

Burning down the Christmas Card List

Monday, December 26, 2005

IT'S BEEN A RECURRING THEME FOR ME IN RECENT YEARS—the idea that we (humans) are naturally programmed to look for transaction-cost optimization, which naturally leads to large batch sizes and "waterfall" thinking.

This year, like any other year, we spend many an evening in December working on our Christmas card list. The first task is to update the information. The data is perishable. People move. Relationships change. Names change. People we haven't heard from for years get dropped from the list; new names are added. My wife and I have lived around the world in the past 15 years. Between us, we've managed to live in Oxford, Toronto, Hong Kong, Singapore, Dublin, Dallas (Texas), Overland Park (Kansas), and Seattle, besides our native Scotland and Japan. We've picked up a lot of friends and colleagues with whom we keep in touch—mostly just that one time a year.

It is interesting to compare my process with my wife's. After gathering an updated list, together with a set of cards, a set of photos of the kids, and stamps, I dutifully write a single card, address the envelope, insert the photo, seal the envelope, and stamp it for sending. In a single sitting I might write ten cards. Cards seldom depart without a personalized, hand-written note inside. It takes effort; I manage to find the energy about two nights per week. My cards are sent off in batches of about ten, perhaps twice a week, starting in early December. I concentrate on the overseas ones and those that require the most detailed notes first. My wife has the same labor input and same quality of finish, but she first addresses all the envelopes, then she writes every single card, then she adds the photos, then she stamps them, and finally, sends them in one single batch. Her large batch of cards is dispatched toward the third week in December, with some doubt as to whether the longer-traveling cards will make it before Christmas. The saving grace is that few of the recipients are from a Christian tradition, and hence, they have a tolerance to receive the cards before or around the New Year rather than by December 25.

My wife's efficiency-driven preference—you have to know her and her focus on economics—leads to a pure "waterfall," large-batch transfer system, with all the value delivered in one large batch at the end, but with additional risk of late delivery to some of our friends. My iterative, flow-of-value approach gets some cards on their way as early as the first week in December. Risk is mitigated. However, it is possible that my iterative, small batch size approach involves more energy expenditure. Regardless, all my cards made it to their destinations before December 25. And, had I run out of energy mid-month (my job at Microsoft has been busy recently, and the commute long and tiresome), at least half my list (those farthest away, and generally, my more important family and old friends) would have received their cards.

Is my approach counter-intuitive? Or is the large batch size, more efficient approach (batch size over transaction cost) the natural way for the human race?

CHAPTER
13

Lean

Recently, a German business partner told me that he'd overheard an Agile coach at a conference say, "What David Anderson has done with Kanban is—he has saved Lean Software Development [from obscurity]!"

Ironic, then, that it is a matter of some frustration to me that a Lean Software Development movement has yet to emerge. The term "Lean Software Development" gets used a lot, but to be honest, there isn't much evidence that Lean concepts actually are being applied to the decision making involved in designing and testing software code.

There have definitely been writings on Lean Software Development—most notably, James Sutton and Peter Middleton's *Lean Software Strategies* (Productivity Press, 2005). And there have been some blogs and presentations from leading Agile people, such as Joshua Kerievsky with his Intentional Design concept, and from James Shore and Arlo Belshee with their Lean Software Workcell concept[1].

However, most of the Lean thought leadership has really focused on management rather than the how-to of doing software. I'm curious why a movement hasn't emerged. There are definitely good ideas around that echo the work of Reinertsen

---

1. "Single-Piece-Flow in Kanban." Lean Software & Systems Conference 2010, Atlanta, http://www.infoq.com/presentations/Single-Piece-Flow-Kanban

in Lean Product Development, such as Software Product Lines (a.k.a. Software Factories), and deferred decision making, as well as option theory in software architecture and design—see my own article on componentization of a domain model at the end of this chapter (page xx). Despite the fact that recognizably Lean approaches to software development exist, and that these approaches are distinct from Agile software development, no movement has formed to drive their adoption. Ultimately, I feel that Lean Software Development is an idea whose time has not yet come! And, by inference, my work with Kanban hasn't saved it. Lean Software Development may yet disappear into obscurity—just another buzzword in the footnotes in the evolution of our profession.

Where have you gone Martin Fowler?
An industry turns its lonely
eyes to you (woo, woo, woo)

Friday, January 14, 2005

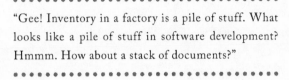

• •

"Gee! Inventory in a factory is a pile of stuff. What
looks like a pile of stuff in software development?
Hmmm. How about a stack of documents?"

• •

IT TOOK A PACK OF LAWYERS FOR SIMON AND GARFUNKEL TO PERSUADE JOE
DiMaggio that they weren't criticizing him, or suggesting that he'd disappeared
from public sight, or that the shine had come off the public image of the man
who'd married Marilyn Monroe. In fact, as the lawyers explained, Joe had to get
used to the idea that he was being used as a metaphor.

Recently, Martin Fowler has been offering us advice about how dangerous
metaphors can be. They are confusing. The problem is two-fold. First, they don't
stretch well (as Alan Cooper famously explained[1]); and secondly, not everyone
can see the abstraction and subsequent concrete example from another field. Joe
DiMaggio was a metaphor for the all-American hero. Here, Martin offers us a
warning about Lean as the all-Agile productivity hero.

> As regular readers of my work may know, I'm very suspicious of using
> metaphors of other professions to reason about software development.
> In particular, I believe the engineering metaphor has done our profession
> damage—in that it has encouraged the notion of separating design from
> construction.
>
> As I was hanging around our London office, this issue came up
> in the context of Lean Manufacturing, a metaphor that's used quite
> often in agile circles—particularly by the Poppendiecks. If I don't like
> metaphoric reasoning from civil engineering, do I like it more from lean
> manufacturing?

1. http://tafein2009.wordpress.com/2009/09/26/the-myth-of-metaphor-alan-cooper/

I think the same dangers apply, but it all comes down to how you use the metaphor. Comparing to another activity is useful if it helps you formulate questions, it's dangerous when you use it to justify answers.

So as an example—one of the principles of lean manufacturing is the elimination of inventory. This leads to the question of whether there is an analogous item to inventory in software development. People have suggested that up front documentation is such an analog. It sits there, producing no value, until you actually deliver some software that's based on it.

Here the metaphor is helping us look at our practices from a different point of view. It helps us to ask questions about what we do. Thus far I think a metaphor is useful.

The breaking point comes when people say: "we eliminate inventory in lean manufacturing, up front documentation is the equivalent of inventory, therefore we eliminate up front documentation". Now I agree we need to substantially reduce this kind of speculative documentation; but the rationale for doing so must come from thinking about the software development process, not from purely reasoning by analogy. —Martin Fowler, Blog, December 16, 2004

Where I want to differ with Martin's view is in his characterization of Lean as a metaphor. Lean Thinking offers us a set of principles. Principles are abstract concepts rather than concrete ones. The all-American hero is an abstract concept. Joe DiMaggio, on the other hand, was a decidedly real example of a baseball player. The English language offers us three words for mapping concepts like this, which, from strongest to weakest, are: abstraction, analogy, and metaphor. Lean is an abstraction rather than a metaphor. Martin Fowler made his name teaching us about abstraction in his seminal work, *Analysis Patterns* (Addison-Wesley, 1996).

At the heart of Martin's complaint with the Poppendiecks' promotion of Lean is the suggestion that "inventory" in software development manifests itself as documentation. As zero (or very low) inventory is a goal in manufacturing, zero documentation should be a goal in software development. Hmmm. Sounds metaphorical to me. So, let's examine this. First off, inventory is an abstract concept that has a specific example in Lean Manufacturing, confusingly, often called inventory, or goods-on-hand. Inventory comes in three forms: raw material, work-in-progress, and finished goods. All three of these terms can apply to the abstract or the concrete. So, it's tricky! The Poppendiecks, thinking in metaphors (rather than abstractions), said, in effect, "Gee! Inventory in a factory is a pile of stuff. What looks like a pile of stuff in software development? Hmmm. How about a stack of documents?" It didn't take long for a few of us to correct this notion, and I notice that recently in the Lean Development Yahoo! group, no one has been suggesting that inventory is documentation. So, Martin is behind the times, though doubtless countless others who've read some of the Poppendieck articles or book will have learned the same notion—that inventory is documentation.

Jason Yip expressed his thoughts via his own blog on why he believes the Lean metaphor is useful:

> My education is in electrical engineering though I did spend some time in graduate school on software engineering. I don't consider engineering software to be a metaphor; I consider what we do to be software engineering. I consider the problem to be people too quickly trying to apply concepts from other types of engineering, usually civil, but even traditional electrical engineering, to software.
>
> The key thing I want to take from the engineering profession is not necessarily any particular technique but rather the "profession" part (i.e., The Obligation and the mandatory internship period).
>
> Every so often on the mailing lists, there is talk about moving toward a apprentice, journeyman, master model for software. Compare this to Engineer-In-Training, Junior Engineer, Intermediate Engineer, Senior Engineer.
>
> In terms of lean manufacturing, of which I am not as familiar, I also question a simplistic application of the metaphor. Inventory in a manufacturing context is expensive because of storage costs. If I just say speculative documentation is bad because it's inventory, the expected counter will be to say that the storage of electronic documentation is negligible.
>
> —Jason Yip, Blog, December 17, 2004

And Clarke Ching had this to say:

> When you discuss lean in your latest blog entry you say that "People have suggested that up front documentation" is an analog for inventory. That is a valid perspective, but it only tells part of the story - and not the most interesting part!
>
> Instead, think of inventory as ALL of the work flowing through the system but not yet sold. Documentation is only a part of inventory. The lean perspective would be that some (most?) documentation is a *waste* because it doesn't *add value* to the customer.

So, there are two types of inventory that lean tries to eliminate:

1) Work in process (WIP)

Traditional manufacturing and Waterfall both *push* work into their systems to keep everyone busy. The end result is that there is a lot of low priority WIP slowing down the processing of the important stuff and lots of waiting time (when they aren't being worked on).

Lean minimises WIP by using a *pull* system and *working in small batches*.

Agile environments minimise WIP by using PULL and *working in small increments*

Both have much less WIP and consequently much shorter lead times.

[They also find defects/bugs much quicker too and have much better visibility and are able to respond to change much more easily.]

2) Unsold Finished goods

Non-lean factories produce goods based on distant *forecasts*. Some parts of the forecasts are wrong so these non lean factories end up producing goods that they either can't sell or they can sell but only at huge discounts.

In Waterfall environments, requirements document are distant *forecast* (guess) too. Some parts of the forecast are wrong so these projects end up producing features that are either not wanted or have to be expensively reworked.

Lean and agile both minimises *unsold Finished goods* by *pulling* work through their systems based on actual demand."

—Excerpt from an email by Clarke Ching to Martin Fowler
published on his blog, December 17, 2004

All of this begs the question from the title . . .

Where have you gone Martin Fowler (international man of abstraction)? An industry still turns its lonely eyes to you . . .

Providing Value with Lean

Monday, March 31, 2008

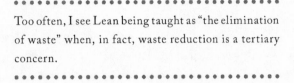

Too often, I see Lean being taught as "the elimination of waste" when, in fact, waste reduction is a tertiary concern.

THERE ARE MANY LESSONS IN THE LEAN PAPER THAT I CO-AUTHORED WITH Merwan Mehta and David Raffo ("Providing Value to Customers in Software Development through Lean Principles"),[1] but perhaps the most important is this message:

❑ Value first, then flow, then waste reduction/elimination

Too often, I see Lean being taught as "the elimination of waste" when, in fact, waste reduction is a tertiary concern. Although waste elimination is important, waste should not be reduced to the detriment of flow, and smooth flow can be sacrificed to improve value delivery.

So, for example, a queue in a kanban system can be considered as waste (probably necessary waste). As queues are a type of waste, it makes sense to reduce or eliminate queues. I see this advice coming from others who talk about Lean in software development.

However, it is important to understand why the queue is there. It is there to absorb variation in size and complexity of work items, or in the availability of someone to process those work items. If you reduce the size of the queue, or eliminate it altogether, it might impede the smooth flow of the system, causing a stop-and-go effect, lowering the overall throughput of the process.

For example (and I talk about this when I present Kanban at conferences), at Corbis we had a non-instant availability bottleneck in build engineering, due to the time-slicing (multi-tasking) required of the build engineers. As a result, we increased the size of the queue in front of build engineering in order to smooth the flow through the whole system. In this example, the variability comes from the availability (or lack thereof) of the resource, not from the sizing of the work

---

1. Merwan Mehta, David Anderson, David Raffo: "Providing value to customers in software development through lean principles." *Software Process: Improvement and Practice* 13(1): 101-109 (2008)

items. So, we increased the amount of "waste" in the system to smooth flow and increase throughput.

It is also known that expediting is bad. Expediting impedes flow and causes WIP to increase and lead times to lengthen. We have demonstrated this empirically with the kanban implementations we've done so far. If we are to focus on smooth flow, we would never expedite.

However, this as a general rule would be wrong. Occasionally, there are times when an expedite request carries a very high monetary value. The impact on flow, and WIP, and lead times to other items in progress is outweighed by the monetary benefit of accepting the expedite request. Hence, it makes sense to pursue the value in the expedite request despite its impact on flow and its introduction of waste into the system.

Hence, value trumps flow, and flow trumps waste elimination in all process-operation and process-improvement decisions.*

● ● ● ● ● ● ● ● ● ● ● ● ● ● ● ● ●

* This has become known as the "Lean Decision Filter" and augmented with a third axiom, "Eliminate waste to improve efficiency, do not pursue economy of scale."

End-to-end Traceability

Monday, April 12, 2004

● ●

"As an Agile guy, I'm surprised to hear you advocating end-to-end traceability. That seems the antithesis of light weight to me."

● ●

WHILE I WAS WRITING MY FIRST BOOK, AND PUBLISHING THE DRAFTS ON THE web for review, I got some feedback from a reader who said, "As an Agile guy, I'm surprised to hear you advocating end-to-end traceability. That seems the antithesis of light weight to me."

I realized that end-to-end traceability meant something very different in this reader's mind than it does in mine. In fact, anyone from the defense industry probably thinks of end-to-end and documentation in the same thought. The term "end-to-end" really relates to the audit trail of documentation—who touched it, when it was modified or updated, and how one document relates to another.

When I'm talking about end-to-end, I'm talking about being able to track value throughout the process. End-to-end means being able to identify the feature a developer is coding or one that a tester is testing and to unambiguously trace that back to a requirements definition in a Product Requirement Document, a Request for Proposal, or some analysis artifact, such as a Use Case. The purpose is to communicate the potential value being created in the system of software engineering such that the consumer, the customer, or the product visionary can know how much of the desired product or service exists at any given point.

I track the flow of value with cumulative flow diagrams (CFDs). It's important to be able to relate the inventory in the CFD to the other artifacts in the project so that the data can be communicated project wide and not just to the development team. It's not important for value tracking to have the audit trail of documents, but it is important to know the milestone progress of each unit of inventory derived out of the initial requirements.

Using Cumulative Flow Diagrams

May 2004

•••••••••••••••••••••••••••••••••••••••

The faster the requirements can be realized as working code and brought to market, the more value will be delivered.

•••••••••••••••••••••••••••••••••••••••

Introduction

Agile software development methods such as Scrum and Feature-Driven Development manage and report project progress in a very different manner than traditional critical-path project management. Scrum started with the use of a "burn down" chart, which plotted the estimated number of hours remaining to complete the sprint backlog. More recently "burn up" charts have become popular. These plot the number of completed Stories, Tasks, or Features on the project with a projected completion target. It is possible to extrapolate the plot with a trend line to estimate the completion date of a project.

The concept of a "burn up" chart had been used in Feature-Driven Development since its inception in the late 1990s. It's called a Feature Complete Graph, as shown in Figure 13.1

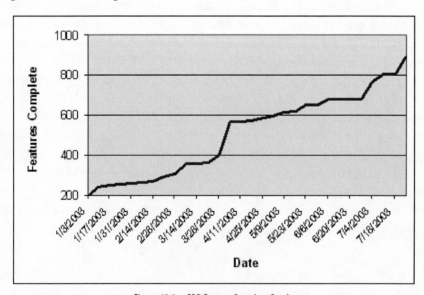

Figure 13.1   FDD Feature Complete Graph

However, "burn up" charts aren't really sufficient for managing a project. They have no concept of work-in-progress (WIP), and simulating the anticipated end date is problematic. In *Agile Management* [Anderson 2003], I introduced Cumulative Flow Diagrams as a better replacement. This article explains why.

The S-Curve

The completion of Features tends to follow an S-Curve model, as shown in Figure 13.2. The S-curve effect makes it difficult to predict the end-date of a project from a single plot of Features complete.

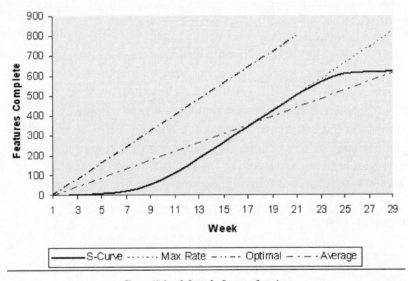

Figure 13.2   S-Curve for Features Complete

In order to communicate a fuller picture of a project's health, I found it necessary to supplement a Feature Complete Graph with a report of the percentage complete and an accompanying graph. A percentage-complete plot for FDD gives credit to Features that are in-progress. Each of the six milestones for a Feature is credited with an earned-value percentage, as shown in Figure 13.3.

Domain Walkthrough	Design	Design Inspection	Coding	Code Inspection	Promote To Build
1%	40%	3%	45%	10%	1%

Figure 13.3   Feature Milestones and earned-value percentages

So, Feature Complete graphs fail to communicate work-in-progress but the Percentage Complete figure and graph suffer from "false reporting." Why? Because any agile developer will tell you that they only value finished, working software. There is no value in partially complete code. This result is compatible with the Theory of Constraints management accounting method, *Throughput Accounting* [Corbett, 1997]. Value should be recognized only on delivery.

Hence, it was necessary to stop reporting percentage complete altogether. Only report the true earned value—Features complete—or find a better way to communicate the work-in-progress, that is, a method that did not report the value of WIP, but did communicate that work was progressing, even when Features were not being completed every day.

Work-in-progress

If it is wrong to communicate earned value from partially complete WIP, then should we ignore it? No, I don't believe so! Rather, we should care about it deeply. Why?

Work-in-progress can be thought of as inventory. In this case, it is knowledge-work inventory—ideas for valuable working software, captured at some stage in a transformational process that takes it through transformations such as analysis, design, test plan, code, unit test, and code inspection (depending on the method you are using). These ideas are developed from the work of Donald Reinertsen, but to understand that, we must first understand the contribution of Marvin Patterson.

Patterson's Design Model

With *Accelerating Innovation* [1993] Marvin Patterson introduced a concept for modeling the design process. He asked us to envisage that design was a process of information discovery. Before a design is started there is little or no information, perhaps only a vague thought. As the design emerges, there is gradually more and more information and less and less uncertainty until the design is complete.

Mary Poppendieck [2003] suggested that software development can be understood with the same model. In other words, all software development is a "design problem." This would allow us to model the flow of value through a software-engineering system as the gradual reduction of uncertainty and the discovery of more and more detailed information until working code, which passes appropriate quality control tests, is produced.

Design is Perishable

Writing in *Managing the Design Factory* [1997], Donald Reinertsen developed the ideas of Patterson a little further by introducing two concepts. The first observation was that design-in-process inventory could be tracked using

Cumulative Flow Diagrams, as shown in Figure 13.4. CFDs were already in use in Lean Production to track the flow of value through a factory. The second, and perhaps the most valuable insight, was that the value of design information depreciates over time. There are several reasons for this. The main one is that information need only be created once, as the cost of replicating it is near zero. If design is information, the time it takes a competitor to duplicate the design is the time in which the design (and its associated information) has a differentiating value. Design information is also perishable because of possible changes in the marketplace—fashions change, and so do laws, regulations, materials, supply components, distribution networks, and business models. To have real value, a design must be appropriate for its time and it must come to market within its window of appropriateness. If software is design, the same must be true of it. The requirements for a software program must be perishable. Hence, the faster the requirements can be realized as working code and brought to market, the more value will be delivered. Reinertsen's and Poppendieck's observations tie software engineering firmly to the principles of Lean, and the lead time for turning an idea into working software must be critical to the financial success of any software activity.

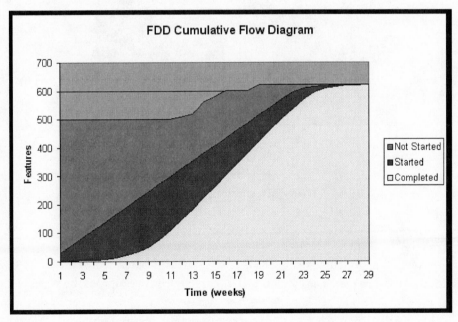

Figure 13.4   An idealized Cumulative Flow Diagram for an FDD project

Little's Law

Little's Law states that a queue of material can be analyzed in two ways: as inventory or as lead time. Simply put, the size of the inventory is directly proportional to the lead time for processing that inventory. Hence, the size of work-in-process inventory matters, because in Agile development we want to complete software in short cycles and deliver value as often as possible. Figure 5 shows how to read the WIP inventory and Lead Times from a CFD.

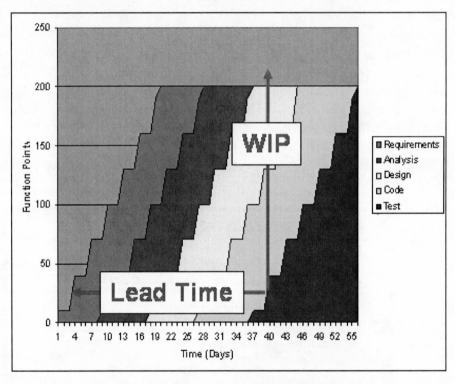

Figure 13.5   Reading WIP and Lead Time from a CFD for Day 40 of a project

Batch Size

Figure 13.5 also demonstrates how batch size and batch transfers affect the cumulative flow plot. The batch transfers can be clearly seen from the jaggedness of the plot. With larger batch sizes, there is more WIP and longer lead times. With smaller batch sizes, as in Figure 13.6, WIP is reduced and lead time falls accordingly. Note the smoothness of the plot in Figure 6. This is from a real FDD project with Chief Programmer Work Packages (small batches of Features) that were completed in less than two weeks. The lead times can be clearly read from the diagram.

It is easy to see from this that CFDs can be used to tell, at a glance, the size of iterations and the type of method being used for development.

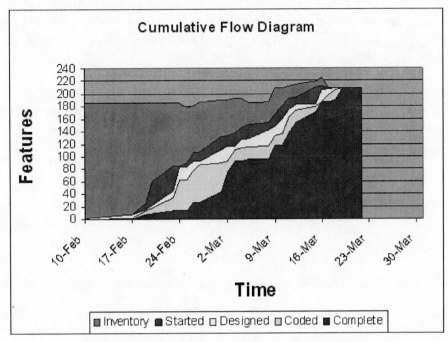

Figure 13.6   CFD showing lead time fall as a result of reduced WIP

WIP is a Leading Measure

Reinertsen describes WIP as a leading metric. What he means is that WIP can predict the lead time and the delivery date. It can therefore be used to correct problems before they become too serious. If we waited to measure lead time or delivery date, there might be greater problems stored up. I once had a project that was reporting 53 percent complete, but only four Features were completed. A CFD plot of this project would have alerted me, as the manager, much earlier. I will discuss how to monitor the health of projects using CFDs in a future Coad Letter.

Summary

Cumulative Flow Diagrams provide a method for tracking progress of agile projects in a "burn up" fashion. Because they plot both the total scope and the progress of individual Features/Stories/Tasks/Functions/Use Cases, they communicate absolute progress while visually providing a proportional message of total completeness. CFDs also offer us a simple method of tracking work-in-progress and visually analyzing the trend in lead time for delivery of working code. They provide a leading metric that allows teams and managers to react early to growing problems and they provide transparency into the whole lifecycle. Tracking a project with a CFD is a key element in moving to a Lean system for software development.

References

[Anderson 2003] Anderson, David J., *Agile Management for Software Engineering—Applying the Theory of Constraints for Business Results*, Prentice Hall, Upper Saddle River NJ, 2003
[Corbett 1997] Corbett, Thomas, *Throughput Accounting*, North River Press, Great Barrington MA, 1997
[Patterson 1993] Patterson, Marvin, *Accelerating Innovation*, Van Nostrand Reinhold, New York NY, 1993
[Poppendieck 2003] Poppendieck, Mary and Tom Poppendieck, *Lean Software Development—An Agile Toolkit*, Addison Wesley, New York NY, 2003
[Reinertsen 1997] Reinertsen, Donald G., *Managing the Design Factory—A Product Developer's Toolkit*, Free Press, New York NY, 1997

Embrace Dark Matter

Monday, February 16, 2004

• •

Why was it better to make a probabilistic guess about the scope and move forward rather than wait to get the analysis perfect?

• •

"DARK MATTER" IS A METAPHOR I USED IN MY FIRST BOOK TO EXPRESS THE IDEA of a requirement that your customer knows to be there, but that you couldn't (or didn't) see when you defined the requirements. Project dark matter is unrecognized, client-valued functionality. Call it "missing analysis," if you prefer. Dark matter represents features that you didn't have on the Feature List, but were there all along. You discover them while doing the detailed design, mid-project. Dark matter causes the feature count on projects to increase. It looks like scope creep to upper management, but the customer denies having asked for any changes.

Dark matter is always there—embrace it!

It is too easy to say, "Next time we will do better analysis," or, "Next time we will spend more time on analysis"—wrong! Embrace dark matter—it's out there! If part of the essence of Agile development is to make progress with imperfect information, dark matter is inevitable. Attempts to prevent it destroy the very agility you seek to achieve.

The real questions are: "How certain are we about this domain?" and "How comfortable are we that we caught all the detail while doing the domain modeling and client-valued functionality identification?" Use this to calculate a scope buffer. On a Feature List, I enter this as empty line items and comment them as "dark matter." During the project, as new features are discovered through more detailed analysis, I replace an empty line with the new feature.

On a recent project iteration, I asked for a 100 percent buffer (that is, double the time) for dark matter due to domain uncertainty. The request was denied. The project ran with a 50 percent buffer (that is, the management contingency was one-third of the total timeline). During the project, 61 features grew to 117 without any change requests from the customer. That's a 92 percent increase due to dark matter. Luckily, we made the date by surprising ourselves and outperforming on both production rate and quality.

Why was it better to make a probabilistic guess about the scope and move forward rather than wait to get the analysis perfect?

Simply put, it would have taken weeks to resolve the analysis issues and gain a high level of certainty that everything had been captured. On the other hand, in four weeks, 117 features were coded and complete with absolute certainty. By embracing dark matter, it was possible to move forward earlier and deliver earlier. In this example, the deliverable was achieved in less time than it would have taken to complete the analysis, had total confidence been a gating prerequisite to moving forward.

Lose the deterministic project management paradigm. Embrace uncertainty using a probabilistic approach to project management. Analysis can never be perfect. When you can accept that, and accept the probabilistic paradigm, you can move forward earlier and complete sooner.

• • • • • • • • • • • • • • • • • •

* This article ties together two very important concepts, one from Agile and one from Lean Product Development. Agile values ask us to make progress with imperfect information and to rework later as we learn more. Lean Product Development teaches us to understand the cost of delay and to recognize that earlier, faster progress to a partial solution is often better than delaying to acquire better information leading to a more complete solution.

These concepts have become core ideas in the Kanban Method and are well understood in the Lean and Kanban communities. Understanding, visualizing, and quantifying the effects of rework (iterative processes) versus cost of delay is both a common goal in Kanban system design and a core fitness criteria for the selection of evolutions of a software development process driven by Kanban.

Revealing Dark Matter

Friday, May 21, 2004

• •

Dark matter is not scope creep, a.k.a. change requests.

• •

I'VE TALKED ABOUT "DARK MATTER" BEFORE—EMERGENT FEATURES THAT ARE missed in the initial backlog analysis but are discovered as the team gets into the design, build, test stage. Dark matter is not scope creep, a.k.a. change requests—those should be plotted differently. Figure 13.7 is an example.

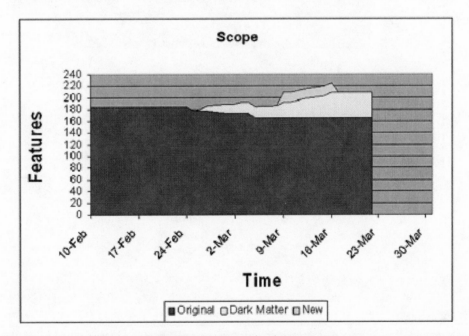

Figure 13.7   Scope creep, a.k.a. change requests

In the chart, initial scope starts to fall as features are pushed from the current iteration to a future iteration. The team gradually discovers that the customer didn't really need those features at this time. Meanwhile, their own analysis gradually reveals the dark matter. These previously hidden features get added to the Feature List and are plotted as part of the Feature Inventory in a Cumulative Flow Diagram (see the CFD for the same project), as shown in

Figure 13.8. When a change request arrives, it is modeled and sized, and it, too, is added to the overall scope. Later, it is discovered that this change isn't really needed until much later, and it is dropped from the Feature Inventory for this iteration, and placed into the future feature backlog.*

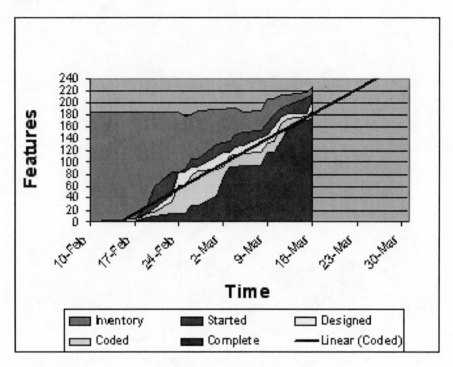

Figure 13.8   Feature Inventory

• • • • • • • • • • • • • • • • •

* Plotting a graph to show the emergent nature of requirements (or demand) is still relatively uncommon. Although it was supported in Microsoft Visual Studio Team System 2005, it has not become a standard report in Kanban software tracking tools. I hope this will change in the next few years. Probabilistically factoring in dark matter expansion to a project plan is a necessary element in developing predictable, long-term plans.

Gondolas and Cable Cars

Sunday, February 13, 2011

• •

As Lean practitioners, we understand that flow systems are more desirable. . . . However, a flow system might not always be the most economically efficient. A discrete batch transfer system might be better for our business, even if customer satisfaction is lower.

• •

TWO WEEKS AGO I WAS RIDING THE AHORNBAHN CABLE CAR[1] FROM ITS BASE station in Mayrhofen, in Austria's Tyrol region, to its summit, 1300 meters above on the Ahornspitze mountain near the Italian border. My long-time friend Karin, a local who was born in the village, was explaining to me the economics of cable cars versus gondolas for transporting skiers to the summit of mountains.

The Ahornbahn

The Ahorn ski area is the much smaller of the two areas accessible from the village of Mayrhofen. It offers a couple of drag tows and two chair lifts, providing access to some easy, wide, blue runs for beginners and a longer, slightly steeper, red run for intermediate skiers. There isn't much variety. As a result, the mountain isn't busy.

However, the Ahorn offers the knowledgeable tourist and the locals the option of a 1300-meter vertical drop back to the base station via the exit run or "abfahrt." The game the locals like to play is "race the cable car." Exit at the top of the mountain, a short walk out of the station to the snow, on with your skis, and off you go. A reasonably fast run is seven minutes. Allowing for the setup time at the top, it is less than ten minutes since the cable car docked and disgorged its load of up to 160 skiers. On a good day, I can finish this run in eight minutes, usually arriving at the bottom just as the cable car returns to the station. My friend is standing there waiting for me with her skis off already.

The Ahornbahn is one of the world's largest cable cars. Mayrhofen has two of the largest in the Alps. The other, holding up to 150 passengers, connects the

1. Americans reading this should translate "cable car" as "tram."

other Mayrhofen ski area of Penken and Horberg with the Rastkogel, above the village of Lanersbach, in the Tux valley.

Most skiers in Mayrhofen access the Penken area via the Penkenbahn gondola. This is a mass-transit system constructed of gondolas hanging on wires approximately 100 meters apart, each holding up to 14 passengers. The Penkenbahn can transport around 2,500 skiers onto the slopes in an hour.

The Penkenbahn gondola

Gondolas move more people per hour; they are more convenient, faster, and provide the skier a better experience than do the cable cars. So, if gondolas are better, why did the Zillertal ski area recently build two of the largest and most sophisticated cable cars ever? Why connect the busy Penken and Horberg areas, with the more difficult red and black runs on the Rastkogel above Lanersbach, with a cable car? Why not a gondola? A gondola would be more convenient for skiers, and would provide a smooth ski-down, ride-immediately-back-up experience.

The explanation is simple—economics!

In Lean terms, gondolas provide a flow system, whereas cable cars provide a batch transfer system.

The flow system of the gondola reduces waiting time and maximizes skiing time. However, a gondola has to be operated constantly, so it costs a lot more to run than a cable car. Since the cable car is a batch transfer system, the batch size is controlled by the size of the cabin.

The lift company estimates demand and forecasts it out over a 10 to 20 year period in the future. They know which days and times are likely to be busier, what skill levels the skiers who are visiting the resort are, and which types of runs they prefer. They know all of this because they collect data from the RFID tag lift pass system.

If they anticipate peak demand at, say, 600 skiers per hour connecting between Horberg and the Rastkogel, then a 150-person cable car operating once every 15 minutes is the economically optimal way to move those skiers.

As Lean practitioners, we understand that flow systems are more desirable. They offer the shortest cycle time, which usually correlates to the highest level of customer satisfaction. However, a flow system might not always be the most economically efficient. A discrete batch transfer system might be better for our business, even if customer satisfaction is lower.*

What if, however, 20 years from now the lift company wants to replace the Rastkogelbahn with a gondola? That would require tearing down the existing system and constructing a whole new one. The project would take nine months, through the spring, summer, and autumn seasons, and involve tens of millions of euros; switching from a batch system to a flow system isn't always easy.

There are many lessons from this example. We should choose batch or flow systems based on demand forecasts. The best choice from the customer's perspective might not be the most economically viable for us as a business. And switching our choice later might be a strategic investment that will take lots of time and money.

● ● ● ● ● ● ● ● ● ● ● ● ● ● ● ● ● ●

* Recently, I was co-training a Kanban class in Munich with Florian Eisenberg of IT-Agile. He had just returned from a snowboarding holiday in the Zillertal. He was unaware of this blog post. Over lunch he was complaining to me about the experience of the Rastkogelbahn cable car. It certainly isn't the best customer experience–but as I'd skied this area for 15 years without a connection between Horberg and Lanersbach, I greatly appreciate that the cable exists at all!

200-Fold Improvement: A Great Yarn

December 28, 2003

• •

Combine the right answer in tools with the knowledge we are gaining about Agile software development . . . and it's a sure thing that we are on the verge of a paradigm-shifting change for the better.

• •

DEVELOPERS AND MANAGERS ALIKE OFTEN LAUGH AT THE SUGGESTION THAT four-fold improvement in programmer productivity is possible, just by changing working practices. Numbers like ten-fold seem like fantasy. Most people I speak to think I'm crazy when I tell them that I believe that 40-fold improvements will be possible within 15 years, and that history will look back on the first 40 years of software engineering as a craft era.

There is a split opinion on the usefulness of history: Although we are often reminded, "Those who do not learn the lessons from history are condemned to repeat it," Henry Ford declared that it was "Bunk!" For those of you who align with the first sentiment, you might enjoy an article entitled "A Great Yarn" [1] from *The Economist Christmas Special*.

The article charts the history of cotton. What has that got to do with software development? In my opinion—a lot! Why?

The Industrial Revolution has a lot in common with the information revolution we are now experiencing. Landed-gentry farmers from England used their wealth to expand into plantations in the colonies. This meant running banana, sugar, and cotton plantations in the Americas, staffed by slaves from Africa. The raw material was brought back to England for added-value processing, and then sold throughout the rapidly expanding British Empire.

Processes like spinning cotton were the high technology of their day. Merchant banking was basically invented during this period to facilitate the flow of capital from the gentry farmers with raw-material wealth to those with ideas for spinning jennies, and steam engines, and locomotive power, and steam-powered looms, and so on, and so on. It was the venture-capital industry of the

---

1. http://www.economist.com/node/2281685/print?Story_ID=2281685

time. A virtuous cycle was started when wealthy people invested in new ideas that generated yet more wealth. *The Economist* does a good job of explaining this for cotton—a key element in the Industrial Revolution, and the (eventual) creation of untold new wealth and higher standards of living for all.

Note how closely the now-unfashionable use of imported slave labor reflects the use today of the H1B and (even more so) L1 visas in Silicon Valley. Rapid expansion fueled by new technology creates labor shortages. Migrant workers fill that demand.

For me, the most important details in the cotton article are the statistics. Over a 70-year period, cotton production got 200 times better. Not only did this not destroy jobs, instead, it created yet greater demand for the product and generated yet more wealth.

I firmly believe that a confluence of two things—management science, which includes knowledge of best working practices, and improved tools—is creating the beginnings of the "spinning jenny effect" for software development. The OMG's MDA; or Steve Mellor's "executable UML"; or Microsoft's Software Factories, based on Software Product Lines research, might not be the right tools, but they are going the right way. They seek to leverage an economy of scope, and they provide us with reusable elements that will multiply the economic effect and move us beyond a purely hand-crafted code industry. Combine the right answer in tools with the knowledge we are gaining about Agile software development and Lean knowledge working, and it's a sure thing that we are on the verge of a paradigm-shifting change for the better. The craft era is ending. If Agile and Lean techniques can produce a 300 to 1000 percent productivity improvement (as has been observed), and new tools for re-use of components, services, and requirements definition produce similar numbers, we are looking at the sorts of economic improvements that manufacturing has realized over the previous centuries. So, who would bet that an article in *The Economist Christmas Special* 70 years from now won't be looking back at changes in software development and knowledge work and reflecting on a new era of economic wealth around the world as a result of a 200-fold improvement?

Centralized Tool Decisions

Thursday, November 6, 2003

●●●●●●●●●●●●●●●●●●●●●●●●●●●●●●●●●●●●●

If you are a development manager, you've probably
had one of your team complain that he's not allowed
to use the best tool for the job because "some idiot at
corporate" forces him to use the standard tool.

●●●●●●●●●●●●●●●●●●●●●●●●●●●●●●●●●●●●●

DOES YOUR ORGANIZATION HAVE A CENTRAL GROUP THAT SELECTS THE TOOLS
that software engineers use? If you work in a large company in the *Fortune* 1000,
chances are that you do. These groups often go by names such as "Technology
Process Group" or "Architecture Committee." Generally, they report to the
corporate HQ on the organization chart. They are usually affiliated with research
groups, and they exhibit the look and feel of an ivory tower.

If you are a development manager, you've probably had one of your team
complain that he's not allowed to use the best tool for the job because "some
idiot at corporate" forces him to use the standard tool. Why does this happen in
software development? In a Lean manufacturing plant, it is unthinkable that a
centralized group of academic manufacturing engineers would make a standard
toolset recommendation. In fact, Lean teams are so empowered that they get to
invent their own specialist tools. Workers are in the best position to decide how
to get the job done. They are expected to choose their own tools for optimal effectiveness. In the production assembly world, where activities are so regulated and
repetitive, this might surprise you. However, when you think about the nature of
software development, it seems downright ridiculous that anyone other than the
software engineer writing the code should choose the tool to get the job done.

So why does this happen—this centralized standard for tools—this proclamation from the ivory tower? Quite simply, it is yet another manifestation of a
cost-accounting fallacy. By choosing a standard tool for, say, object modeling or a
development IDE, there is an economy of scale for a larger license purchase. This
is a savings, right? It reduces operating expense, OE, right? Yes, this is potentially
true, but what does it do to throughput?

In a world where it is now widely recognized that no one software-
engineering process fits all sizes or domains of project, it is also highly unlikely
that one tool fits all types of software development.

There is another kind of argument proffered for single-tool selections: It points to interoperability of personnel and ease of support and maintenance. Again, this is a cost-accounting fallacy. It exists because tools are confused with platforms. Platforms are technology environments that run in production. A RDBMS, such as Oracle, is a platform. An application-server middleware, such as Weblogic, is a platform. A runtime environment, such as Java or .NET, is a platform. However, an IDE, or a modeling tool, or a compiler, or a version-control system is simply a tool. Platforms are involved in generating throughput for the operational business. A platform must be maintained in production. It makes sense to reduce the number of platforms that must be supported. It does not make sense to cripple a development manager's ability to deliver a project by forcing his or her team to use the wrong tool for the job.

Now, to be clear, I am not advocating a free for all. A team that has every member using a different IDE is not a team, but a group of egos flying in close proximity. Teams must decide on the tools to make them effective. This should be done on a per-project basis. The workers writing the code should make these choices by consensus, and the development manager should have the freedom to procure the tools that the team needs. Centralized tools standards represent constraining policies. The Theory of Constraints suggests that policy constraints should be eliminated whenever possible. Centralized policy constraints disempower development managers and developers. This creates a poor environment for optimal productivity. It demoralizes knowledge workers and constrains their performance.

The agile senior executive seeks to eliminate central committees and their cost-accounting-inspired, policy-constraint proclamations whenever and wherever possible.

● ● ● ● ● ● ● ● ● ● ● ● ● ● ● ● ● ●

This is the first of two articles that I wrote on the Lean topic of Standard Work. I believe that although some standards are important, the Standard Work concept from Lean Manufacturing is perhaps the most dangerous concept to translate into knowledge-work fields. I believe it is easier for industrial engineers to determine a best practice or an area of risk where standardization is necessary versus areas where teams can be empowered to develop their own tools. At the time of writing in 2012, I see very little good and useful guidance on Standard Work, while there are numerous anecdotes of bad examples, including, for example, standard desk layouts and positions for keyboard and mouse. It seems to me that there are very few areas of knowledge work where a "best practice" is possible. More likely we have "good practices" or we have emergent responses to complex situations. As a result, standardizing tools as a "best practice" is probably destructive.

Centralized Process-Selection Decisions

Tuesday, November 18, 2003

* *

There should be no grand, centralized choice made
in the ivory tower.

* *

IT IS NOW WIDELY RECOGNIZED THAT WITH SOFTWARE-DEVELOPMENT
processes, one size does not fit all. This goes as much for eXtreme Programming
as it does for a traditional SDLC or RUP. In "The Right Tool for the Job," Scott
Ambler examines this topic in the latest issue of *Software Development* magazine.
Scott references using risk assessment as a tool to help select the right process,
and he points the reader to the recent Boehm and Turner book, *Balancing Agility
and Discipline* (Addison-Wesley, 2003).

Last week, I talked about the problems of having a centralized group that
selects tools for software developers to use. The same applies to software process.
It makes no sense to have a centralized proclamation that the entire enterprise—
and I've worked in companies with 20,000 software developers—should use RUP,
for example. Software-process choices should be aligned with workflows. The
market into which the software is being deployed should be well understood.
Are the requirements stable? Is it likely that the application can be deployed
iteratively, or incrementally, rather than holistically? This is what Boehm and
Turner call a risk assessment. There are other treatments of the problem.

In Chapter 34 of *Agile Management,* I divide the problem into a two-by-
two matrix with four categories: immature holistic domains, mature holistic
domains, immature incremental domains, and mature incremental domains. I
then suggest process choices for each of these quadrants.

In *Agile Software Development Ecosystems* (Addison-Wesley, 2002), Jim
Highsmith maps the problem space using Geoffrey Moore's technology-adop-
tion lifecycle model—Early Market, Chasm, Bowling Alley, Tornado, and Main
Street—as described in his books, *Inside the Tornado (HarperBusiness*, 2004),
and earlier, in *Crossing the Chasm* (HarperBusiness, 1991, 2002).

There is lots of advice out there. The bottom line is that this advice should be
followed on an as-needed basis—by project or by business unit. There should be no
grand, centralized choice made in the ivory tower. The minor cost-efficient advantage of the whole staff being trained in one method is far outweighed by the problems created—and the real cost, in ROI terms—by using the wrong tool for the job.

Go to the Source, or Use
Your Imagination?

Thursday, March 16, 2006

• •

The techniques of Personas, Lifestyle Snapshots, and Usage Scenarios were all developed to allow software people to use their imaginations to envisage real people doing real things with a product that doesn't yet exist.

• •

I BELIEVE THAT THESE IMAGINATION-BASED TECHNIQUES DELIVER MORE THAN an 80–20 payback;, i.e.that is, for 20% percent of the expense of "going to the scene" and observing what's actually happening, we can get 80% percent of the user-experience results by using our imaginations.

I've noticed recently that many in the Lean community have moved their thinking beyond elimination of waste. The two hot topics are *genchi gembutsu* ("go to the source") and set-based design (see later article in this chapter).

The idea of going to the actual scene and experiencing problems for real interests me—because we do it so seldom in software development (particularly in product development.) But we have found a way of compensating: We use our imaginations. The techniques of Personas, Lifestyle Snapshots, and Usage Scenarios were all developed to allow software people to use their imaginations to envisage real people doing real things with a product that doesn't yet exist. I believe that these imagination-based techniques deliver more than an 80–20 payback; that is, for 20 percent of the expense of "going to the scene" and observing what's actually happening, we can get 80 percent of the user-experience results by using our imaginations.

In any event, the idea of "going to the scene" does already exist in our industry's literature. The (misnamed) concept of ethnographic study has existed in usability engineering for 20-plus years. This is misnamed because such studies are truly anthropological studies of specific situations rather than ethnographic studies, but somehow the wrong word stuck and it is in wide usage in the literature.

So, to those who are touting *genchi gembutsu* as the next big thing in Lean software development, I'd reply that it's not a new idea—we've had an equivalent idea for a very long time. However, it is not widely used because it is expensive and difficult to do.* Instead, we have developed an alternative that utilizes our imaginations and artistic flair. I'd argue that we achieve the essence of *genchi gembutsu* in software through imaginative techniques, such as Personas and Usage Scenarios, and not through physical presence at the scene. (See the related post, When Genchi Gembutsu Goes Wrong.)

● ● ● ● ● ● ● ● ● ● ● ● ● ● ● ● ●

* As with most things that are expensive and difficult to do, genchi gembutsu turned out to be a bit of a fad in the community. The Lean Startup ideas from Eric Ries and others that encourage releasing minimalistic or partial solutions to validate assumptions quickly with real customers seem to be a better approach. Don't go to the source beforehand, rather, spend very little time and money and go to the source with something that works and get feedback.

When *Genchi Gembutsu* Goes Wrong!

Friday, March 17, 2006

• •

The anthropological study revealed several aspects
of about how the bankers worked that hadn't weren't
captured by in the business analysts who wrote the
original requirements. This, on the surface, looked
like a flaw in the requirements, and initially I was
heralded as the guy who had uncovered the truth.

• •

IN SINGAPORE, IN 1997, I WAS THE USER-INTERFACE ARCHITECT ON THE FIRST
Feature-Driven Development project, at United Overseas Bank. This is well
known. Much of how I conducted the UI design effort was captured in papers I
published at uidesign.net, but not all of it. It isn't widely known that I went "to
the scene" (*genchi gembutsu*) and sat with the bankers. I watched them as they
worked, and I asked them questions about their work environment and what
they were doing. I built several studies from which we derived requirements for
the system and the user interface.

However, it wasn't a total success. The anthropological study revealed several aspects about how the bankers worked that weren't captured in the business requirements. This, on the surface, looked like a flaw in the requirements,
and initially I was heralded as the guy who had uncovered the truth. However,
it turned out that many of these items were, in fact, aspects of a badly broken
paper-based system. Had they been implemented as I'd recorded them, they
would have led to increased complexity in the system, and they would have
digitally institutionalized inefficiency. The analysts had understood the depth
of the banking system and had designed out these inefficiencies when they'd
written the requirements. As the new guy on the scene, doing surface-deep user-
interface observations, I merely sought to recreate their environment. We had to
go back, rework the analysis, and devise a model that was better and a UI that
would be intuitive. To do that, we had to use our imaginations. Going to the
scene wasn't enough. That produced a local optimization because my visibility
was local to individual users, not system wide. It took me several months to fully
understand the wider banking system.

Set-Based Design

Tuesday, May 2, 2006

• •

The idea with set-based design is that multiple designs are created and kept alive until, gradually, they are eliminated, and a best choice is made.

• •

RECENTLY, THE LEAN COMMUNITY HAS BEEN ENAMORED WITH TWO HOT TOPICS, now that elimination of waste and value-stream mapping are passé. I made my feelings known about *genchi gembutsu* (going to the source) in the previous two articles. The other hot topic is set-based design. The robot above belongs to Katherine Radeka, who was the chair of the recent Lean Design and Development conference. Katherine uses the robot in a new, set-based design theory training class she teaches to (mostly) software people.

Figure 13.9   Robot used to teach set-based theory

The box of parts is used to create alternative architectures for the problem. The original robot is not designed to pick up the axle in the top photo. It has a new problem to solve—pick up the axle and place it between the goal posts. A set of alternatives will be developed and tested for performance, manufacturability, cost, and so forth.

The idea with set-based design is that multiple designs are created and kept alive until, gradually, they are eliminated, and a best choice is made. The best choice comes from a combination of economics, manufacturability, performance, function, and so forth. Set-based design is a logical extension from the Lean concept, "Decide at the last responsible moment." In this case, it is the architecture that is being locked down at the last responsible moment. Actually, manufacturing has another name for this, "postponement." It is the idea that you postpone decisions until as late in the value chain as possible. In some cases, postponement goes all the way to the point of retail sale. Remember removable faceplates for cell phones? Or how about paint that is mixed in the store at the point of sale? Remember when stores carried an inventory of colors and color fashion changed with the years?

Figure 13.10   Parts used to create alternative architectures

The new set-based design hype worries me a little because, for years, we've had a body of knowledge in our business around developing alternative architectures. The Software Engineering Institute offers us the Architecture Tradeoff Analysis Method (ATAM). In my work at Microsoft I picked up work by Jeromy Carriere at Microsoft on Lightweight Alternative Architecture Assessment Method (LAAAM), that offered an potentially Agile alternative to ATAM. The idea that we develop alternative architectures, assess them, and choose the best solution based on a range of criteria is not new. In fact, making architectural alternatives is required for CMMI Level 3 as part of Decision Analysis and Resolution (DAR).

Set-based design is bringing introduces use real-options theory and the postponed decision making. This is not really present in ATAM or LAAAM, which have a single, early-lifecycle decision point.

Despite the hype from the Lean movement, there is a gap. No one with real software-engineering systems and architecture experience has occupied the space and explained how to synthesize postponement and real-options theory with alternative architecture assessment.

●●●●●●●●●●●●●●●●●

* Set-based Design has made little progress in Lean Software Development. I believe the core reason is that software is inherently "soft," and the cost of change is generally lower than the cost of developing parallel alternative solutions. Software can be architected to leave options open and defer design decisions, or decisions can be made and then changed at relatively low cost. For these reasons, Donald Reinertsen believes Set-based Design isn't relevant in software development.

From this series of articles on Standard Work, Genchi Gembutsu and Set-Based Design, it is easy to see that well established Lean techniques from other industries might not make sense in the software-development world. Hence, I remain skeptical of the growing community of authors and thought leaders who seek to map concepts from other domains into software. Until a concept is proven in the field and shown to produce better results, it probably isn't worth taking seriously.

The Lean concepts that have stuck, and are worth pursuing, include flow of work, kanban systems, cost of delay, understanding variability, and deferred commitment (option theory), whereas things like standard work, genchi gembutsu, and set-based design haven't provided much value.

Good versus Bad Variation

Thursday, June 22, 2006

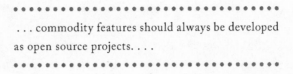

. . . commodity features should always be developed
as open source projects. . . .

GOOD VERSUS BAD VARIATION WAS A THEME AT LAST WEEK'S LEAN MANAGEMENT Summit in Chicago. It has clearly been a focus in Don Reinertsen's[1] recent work. And Brad Appleton[2] was talking about it, too, recently. I've talked about it, too. And then, this morning, Martin Geddes[3] sent me a link to a piece from J. P. Rangaswami[4], which has some good basic advice that sums up the consensus at the Lean Management Summit:

- ❏ *Organizing for routine work: Drive out variation*

- ❏ *Organizing for innovative work: Encourage variation*

The post goes on to grind on some old and outdated views of manufacturing. Surely, manufacturing has a focus on driving out variation. However, it isn't true to claim that assembly lines are only capable of producing Model T Fords in black. In fact, modern manufacturing facilities are very flexible, and modern automobiles are offered in as many as 30 billion configurations. It is, in fact, unusual for an assembly line to turn out more than a handful of cars in a single year that are actually identical.

It was this view of manufacturing that inspired Jack Greenfield in his vision for Software Factories. The idea, that we can capture routine development work and encapsulate it as a reusable package of assets in something we (at Microsoft) call a Software Factory—which can be deployed into an application-development project at minimal cost and effort, enabling the development team to focus on

1. Reinertsen, Donald G., *The Principles of Product Development Flow: 2nd Generation Lean Product Development*, Celeritas Publishing 2010

2. http://bradapp.blogspot.com/2006/05/six-sigma-and-good-vs-bad-variation.html

3. http://www.telepocalypse.net/

4. http://confusedofcalcutta.com/2006/06/22/four-pillars-on-innovation-and-education/

the truly innovative and differentiating features in the design—is one that will greatly improve quality, reliability, and productivity in software engineering.

This concept can be tied to my ideas on strategic planning and product mix selection, which suggest that we can divide a product mix into "table stakes" (or commodities), differentiators, and spoilers. The commodity features are the ones we want to encapsulate in a Software Factory. JP Rangaswami suggests that commodity features should always be developed as open source projects. W see the role of Software Factories as enabling a supply chain for commodity, reusable, application development assets. Software Factories support capitalism and the development of a supply chain in our industry, similar to the supply chains that exist for other industries, like automotive. Sure, some commodity components might be developed as collaborative industry initiatives. Some of those will be open source, but the key here is that it doesn't always have to be that way. There is a capitalist alternative that potentially enables an exchange of value between horizontal market creators of commodity components and vertical, or niche market, operators delivering differentiated products and services.

● ● ● ● ● ● ● ● ● ● ● ● ● ● ● ● ●

* It is a matter of some frustration to me that there hasn't been the emergence of a true Lean software development movement distinctly separate from the Agile software development movement. The Agile movement, with its craftsmanship ethic, has rejected the Software Product Lines/Software Factories' "economy of scope" and reuse of code fragments. This somewhat Luddite behavior is disappointing, and I believe it is holding back a lot of progress. At the time of writing, I am still unclear on how to bootstrap a Lean software development movement. I've tried adding technical-practices tracks to the Lean Software & Systems Conference, but these sessions were poorly attended, and the conference has failed to attract a significant technical audience. Lean software development (after almost 20 years) seems to be an idea whose time is yet to come.

Coarse-Grained Components
from a Color Model

August 13, 2004

• •

One fear held by developers is that if they do not
do a high-level, top-down, component definition,
but simply jump into a wide domain–modeling
exercise with the full scope of requirements, they
will simply be left with one big, monolithic system
that cannot be subdivided.

• •

Introduction[1]

I am often asked, "What is the relevance of domain modeling in an age of distributed application frameworks and service-oriented architecture?" One simple answer is that there is no replacement for good analysis. Domain modeling in color helps you to understand a problem, analyze it thoroughly, and communicate it clearly. It gets everyone on the same page. However, if that were all it did, you would only ever need a simple drawing tool for models and there would be no need for a sophisticated tool that generates code and keeps it synchronized with the model.

Writing in *Bitter EJB*, Tate et al [Tate 2003] demonstrate that good component architectures should have a fully normalized, domain-driven, functional design built within a set of coarse-grained components. These coarse-grained components should exhibit a service-oriented interface, externally enabling a service-oriented architecture. This means that domain modeling in color has relevance for the component definitions in a service-oriented architecture.

Component Definition Ambiguity

There is a tendency, at the beginning of system design, for teams to make a high-level attempt to define a set of components from a sketchy understanding

---

1. This article is archived on the Code Gear site at http://edn.embarcadero.com/article/32541

of the domain. This becomes the outline for the system architecture. Next, in a traditional design approach, the team will try to define interfaces or a set of services to be provided by each component. Once again, this is being done with a sketchy understanding of the domain. Typically, this component definition is done before doing detailed domain modeling using the detailed requirements, and without business owners and subject-matter experts present. In larger companies, often, a different team does it. This is a top-down approach!

Component definitions might later be distributed to geographically dispersed teams who are asked to develop, independently, the interior of each component. This technique of defining interfaces is considered a best practice because it mitigates risk and encapsulates the implementation of individual components. It allows for parallel development with low risk. It is based on the underlying assumption that the interfaces do not need to change and that the component boundaries are correctly drawn in the first instance. This is a typical systems-engineering approach. It, however, produces very brittle architectures that carry significant risk due to a lack of understanding of the domain and the requirements at the beginning

What gets many distributed computing projects into trouble is that these component boundaries are often ill-conceived and they become a burden—a drag on software development and an impediment to reuse, or they require refactoring later. Why is this so?

As each team begins to analyze in-depth requirements for their piece of the system, they struggle with understanding precisely what fits within their scope, which results in debate about the responsibilities of their component. Often, as they gain a deeper and better understanding, they decide it is better to delegate some responsibilities to other components. This results in communication across teams, and, more than likely, a refactoring of the agreed-upon interface. This causes confusion, delay, and it might impact quality. In turn, this causes further delay.

I've had better results with a bottom-up approach to component definition; the following explains how.

Postponement and Lean Software Development

In the manufacturing industry, the concept of postponement is well understood. Postponement means leaving a decision—making it as late as possible. This has been shown, in Lean manufacturing, to minimize inventory levels and reduce lead times, while improving customer satisfaction. Postponement means designing a product such that fickle end-user choices, such as color, or shape, or fabric, or pattern, or user interface—such as language settings—can be committed to as late as possible. No manufacturer wants to be left with a large inventory of purple widgets because they were caught by a consumer sentiment change and purple is suddenly last year's color.

Mary Poppendieck introduced the principles of lean manufacturing to software development [Poppendieck 2003]. One of those principles is "Decide as late as possible," or stated another way, "Keep options open until the last responsible moment." Poppendieck was specifically referring to software lifecycle processes, but the principle can and should be applied to systems architecture. Decisions about architecture deeply affect the likelihood of reuse, of future refactoring, and of flexibility or adaptability. Architectural decisions can and do affect the future velocity of development teams, whether they are using an Agile or a traditional approach. Postponing architectural decisions to the last responsible moment allows for flexibility, adaptability, and responsiveness, but most of all, it allows the maximum time to acquire the most information about the system being built. More information means less uncertainty, and less uncertainty means less likelihood of change becoming necessary later.

The DNC facilitates postponement of component boundaries

One fear held by developers is that if they do not do a high-level, top-down component definition, but simply jump into a wide domain–modeling exercise with the full scope of requirements, they will simply be left with one big, monolithic system that cannot be subdivided. The fear is that there will never be time for the required massive restructuring of such a monolithic system into more flexible components. There are plenty of real-world case studies to underscore the reality behind this fear.

However, the Domain Neutral Component (DNC) and the color modeling technique are designed for optimal, loosely coupled systems that can be easily componentized. With the DNC, every class is treated as a component, and every method on every class is a service being provided by that class. By applying the "Law of Demeter" [Anderson 2004], each class holds dependencies only to its immediate neighbors. In a typical DNC model, that means that most classes hold relationships to only two or three others. This means that it is easy to defer definition of coarse-grained component boundaries until later. All that is required is minor-impact refactoring to resolve any two-way dependencies. The interface of the coarse-grained component will inherit the methods from the classes on its boundary, which are available to those classes that now reside in a different component. Figure 13.11 shows the guidelines for packaging, or componentizing, a DNC model.

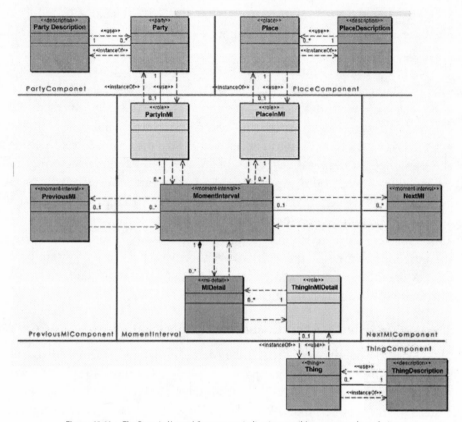

Figure 13.11 The Domain Neutral Component, indicating possible component boundaries

Figure 13.12 shows how this would look as a UML Component diagram after any two-way dependencies have been resolved. Note that the dependency direction is from the transactional components—which contain the <<Moment-Interval>> and the <<Role>> classes of the <<Party>>, <<Place>>, or <<Thing>> components—on which it relies to complete a transaction. Any two-way dependencies that might emerge from a sequence of transactional <<Moment-Interval>> components can be resolved with the inclusion of a <<Factory>> component, which makes (constructs) the subsequent (or next) <<Moment-Interval>> in the chain.

There are three other relatively likely schemes for packaging/componentization. One is to make a component from each layer of a DNC model, that is, package the whole <<Moment-Interval>> graph, with its associated Parties, Places, and Things. However, this strategy is risky because it denies easy reuse of the Parties, Places, or Things to other components. Use this strategy only when you are certain about the domain and the limited need for reuse of the green classes within this DNC graph.

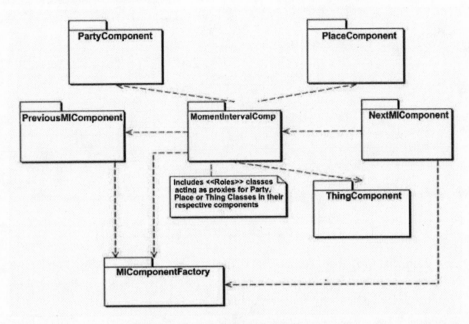

Figure 13.12   DNC showing Dynamic Dependencies

Figure 13.13 shows another, more granular strategy, which separates out across the whole-part relationship between a <<Moment-Interval>> and its <<MI-Detail>>. You would use this strategy where the <<MI-Detail>> needs to be reused across generations or versions of its containing <<Moment-Interval>>, or the <<Moment-Interval>> and <<MI-Detail>> need to vary independently for some other reason. This might be common where the M-I and its detail are controlled by two different industry specifications that develop independently. I have seen this pattern in telecom applications.

Figure 13.14 shows another approach, in which the entire chain of <<Moment-Intervals>> in a workflow is encapsulated as a single component. Note that all the <<Role>> classes go inside this component. This pattern would be appropriate where the workflow is well understood or where it is likely to vary as a whole, and the individual steps do not need to be reused independently.

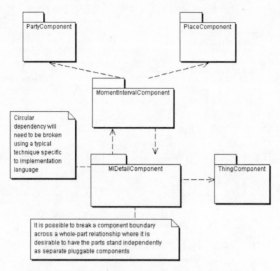

Figure 13.13　MI-Detail is separated as its own component

Note that in all three component diagrams, the dependency direction is from the component with the <<Moment-Interval>> and the <<Role>> out toward the components that hold the green and blue classes. This makes sense because, ultimately, it is the <<Party>>, <<Place>>, or <<Thing>> classes that we want to reuse in other applications and systems. Their respective <<Description>> classes may also be reusable or extendable through <<plug-in>> points to cope with flexible reuse. I will discuss plug-in points in another *Coad Letter* later this year.

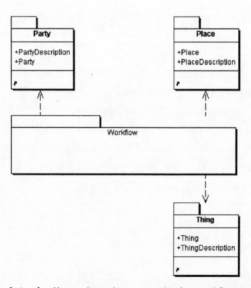

Figure 13.14　Series of <<Moment-Intervals>> encapsulated as a workflow component

Resolving Two-way Dependencies

Figure 13.15 shows a generic approach to resolving two-way dependencies in a color model for the simple task of packaging in Java. This example shows how to resolve a dependency across a whole-part relationship by introducing a collector interface. Figure 13.16 provides a domain-specific example from a hotel and event management application. The ICollectSession interface is placed inside a common package, and both the conference package, which contains the Conference class, and the session package, which contains the Session class, have a one-way dependency down to the common package that contains the ICollectSession interface. This allows libraries to be created from the classes in a color model and, consequently, reduces compile times.

Despite the recent trend to continuous integration, I still strongly believe in packaging and library usage. Using libraries encourages developers to think carefully about responsibilities, and to clearly identify when a class has been affected by a change. If a library needs to be recompiled, it clearly needs to be regression tested. If it is unchanged, it need not be tested. The use of libraries encourages cleaner, loosely coupled code, which, in turn, often leads to higher quality and greater reuse.

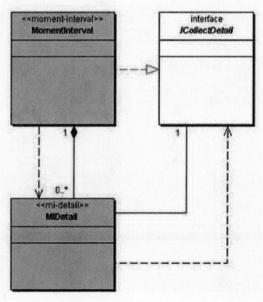

Figure 13.15. Generic approach to resolving two-way dependency in Java

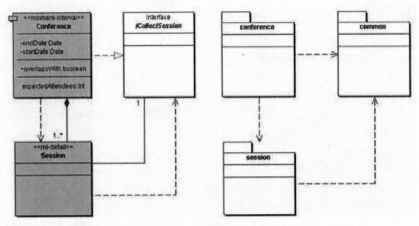

Figure 13.16 Eliminating a two-way dependency with a common package

Figure 13.17 shows a similar technique, employed between a <<Role>> and its <<Place>>. Note how the interface defines the methods that the green class needs to call on its yellow <<Role>> class. Only the methods called across the boundary of the package should be exposed in the interface, but all of the methods required must be exposed in the interface. Again, the interface would be packaged in the common package, as shown in Figure 13.18.

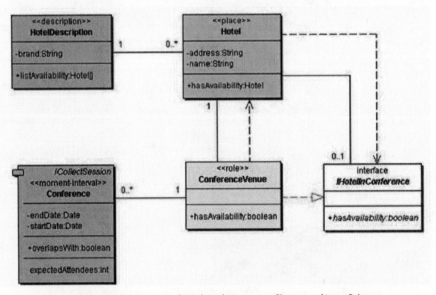

Figure 13.17 Resolving two-way dependency between a <<Place>> and its <<Role>>

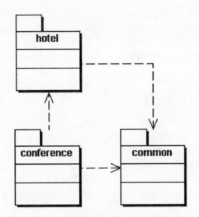

Figure 13.18 Package dependencies for Figure 13.17

Summary

Domain modeling in color with the Domain Neutral Component still has relevance in a world of component frameworks and application servers. Domain modeling allows for precise discovery of a problem domain, and this enhances communal understanding. It provides the ability to postpone coarse-grained component definition without the risk of monolithic, unmaintainable system architecture emerging. Because each class in a DNC model is treated as a component in its own right, aggregation of components is a natural thing that provides for better, more accurate assignment of responsibilities, and it reduces or eliminates refactoring of the coarse-grained component boundary and its interface.

Using a color model and the DNC pattern as a foundation for component framework architecture leads to faster development of the system architecture, reduced rework, better intra-team communication, and, ultimately, more reusable, loosely coupled components. Technical practices such as these are core to delivering a vision of Lean Software Development.

References

Anderson, David J., "Color Models and the Law of Demeter," *The Coad Letter,* Borland Developer Network, Scott's Valley CA, 2004

Coad Peter, Eric Le Febvre, and Jeff De Luca, *Java Modeling in Color with UML—Enterprise Components and Process,* Prentice Hall, Upper Saddle River, NJ, 1999

Palmer, Stephen R., "A New Beginning," *The Coad Letter,* Borland Developer Network, 2002 http://www.thecoadletter.com/article/0,1410,29697,00.html

Poppendieck, Mary and Tom Poppendieck, *Lean Software Development—an Agile Toolkit,* Addison-Wesley, New York, NY, 2003

Tate, Bruce, Mike Clark, Bob Lee, and Patrick Linskey, *Bitter EJB,* Manning Publications, 2003

Contingency Planning

Wednesday, October 25, 2006

I left the house without my underwear this morning!

It's not unusual that I'm not wearing any underwear. Three or four days per week, I bike to work, and the bike shorts suffice. However, my underwear is usually safely stowed in my messenger bag, nestled in with my trousers and shirt in the main compartment, separated from my office laptop. Safely stored, ready for use when I emerge from the office shower room 30 minutes later.

However, today was different. I was halfway to work, cranking a big gear down Fifteenth Avenue, just passing the city pound (ahem, animal shelter) when it hit me. Oh no, I forgot to pack something. Aaaaaggggghhhhh! Never mind, I had traffic to worry about. I'd made it to Second Avenue in Belltown, about five minutes from my office, when I suddenly remembered my contingency plan. I had a packet of spare clothes stuffed in the back of my filing cabinet. In fact, this contingency plan had been in place for two years already, and the packet of clothes had moved offices with me from Microsoft to Corbis recently. The question was, "Is there a pair of underpants in the packet?"

The original risk was identified as "What if I get soaked cycling to work and the clothes inside my rucksack get wet?" The mitigation was to store a spare set of clothes at the office. The contingency plan was to use those clothes in the event of a drenching. Now, despite Seattle's reputation as Rain City, it doesn't actually rain all that often. Indeed, it was dry today. But I made a mistake; I was too sleepy. And although I remembered to pack my workout clothes for my personal training session at the gym beside the office, I forgot something small but vital. Luckily, I was able to invoke the contingency plan for a similar risk with the same outcome.

Risk management doesn't need to be difficult. It's just a matter of thinking through likely outcomes—errors that can occur, external variations out of your immediate control—and then mitigating them by putting in place suitable recovery plans, contingencies. Luckily for me, forethought two years earlier saved my blushes.

Now I just need to remember to replenish the filing cabinet before it happens again. ;-)

Random Thoughts on Risk

Pᴿᴼᴶᴇᴄᴛ ᴍᴀɴᴀɢᴇʀs ᴏғᴛᴇɴ ʜᴀᴠᴇ ᴀ ғᴀɪʀʟʏ ɴᴀʀʀᴏᴡ ᴅᴇғɪɴɪᴛɪᴏɴ ᴏғ ʀɪsᴋ ᴍᴀɴᴀɢᴇᴍᴇɴᴛ, ʙᴀsᴇᴅ ᴏɴ ᴡʜᴀᴛ ᴛʜᴇʏ'ᴠᴇ ʀᴇᴀᴅ ᴏʀ ʟᴇᴀʀɴᴇᴅ ɪɴ ᴏʀᴅᴇʀ ᴛᴏ ᴘᴀss ᴛʜᴇɪʀ Pʀᴏᴊᴇᴄᴛ Mᴀɴᴀɢᴇᴍᴇɴᴛ Pʀᴏғᴇssɪᴏɴᴀʟ (PMP) ᴇxᴀᴍɪɴᴀᴛɪᴏɴ. Pʀᴏᴄᴇss ᴇɴɢɪɴᴇᴇʀs ғᴀᴍɪʟɪᴀʀ ᴡɪᴛʜ ᴍᴏᴅᴇʟs sᴜᴄʜ ᴀs ᴛʜᴇ Capability Maturity Model Integration (CMMI) often share a similarly narrow view—Risk Management is a single process area in the CMMI model. Both of these examples focus on "event-driven risk." Event-driven risk is the concept that some event that would hinder or affect the successful completion of a project or product development has a likelihood of occurrence, an impact on occurrence, and can be managed through efforts to reduce the likelihood or mitigate the impact. If the event occurs, some contingency might have been made to recover from the impact or to implement an alternative approach (a "plan B").

Although this is a perfectly reasonable definition for risk management, and the body of knowledge on this topic is, in my opinion, of good quality, pragmatic, actionable, and effective, my use of the term "risk" in the title of this chapter has a much broader meaning.

"Risk," in this sense, refers to the uncertainty being managed by the enterprise as whole. It doesn't just refer to specific events, but to the broad uncertainties in the market. Risk includes elements such as wider economic factors; the competitive marketplace; the quality of our product; how well our prospective customer likes our product, and how quickly they adopt it; our ability to understand our market, and to segment it appropriately; our capability to deliver against fluctuating demand, and our ability to respond to unfolding events in a timely manner; our propensity for innovation; and our strategic and market positioning. Risk management, in other words, is everything we do, in a specific context, to make our business activities successful. It is the essence of knowledge work—as Drucker described, it's the executive decision making that affects the bottom line of our business. All knowledge workers are, by implication, risk managers.

Risk management is clearly too big a topic for a mere chapter in a book on management. It deserves an entire book, perhaps a series.

The Agile movement has lacked a true sense of risk management. The notable authors on the topic are Alistair Cockburn, Barry Boehm, and Richard Turner. Others in the community who've written and presented on the topic include Todd Little, Kent McDonald, Niel Nickolaisen, Olav Maassen, and Chris Matts. This is worth mentioning if only to highlight how few people have been willing to address wider risk management within the Agile space. A Google search for Agile Risk Management produces a few links to a small-batch size, incremental, time-boxed approach to traditional project management event-driven risk management.

The originator of the term, "Lean Software Development," Robert (Bob) Charette, told me that he discovered Lean and pursued it in software development as a means to pursue better risk management. Bob sees Lean as a core enabler of improved risk management. Techniques within Lean, such as Kanban, can help to expose the risks being managed, and in so doing, improve their management. That's a topic for a future book. In the meantime, here are some of my collected thoughts on how knowledge-worker executives, such as software developers, can better manage risk for their businesses.

Hanami Planning is Non-Deterministic

Sunday, April 10, 2005

• •

In our real, home lives, we know how to deal with variation. We know how to set expectations appropriately. We accommodate variation as if it were natural, because it is. Why do we struggle to do so in our professional lives?

• •

THE JAPANESE SEASON OF *HANAMI* (FLOWER PARTY) IS TAKING PLACE IN TOKYO this week. From a management perspective, this means an excuse to organize a morale-boosting party for your staff. Here's how to do it. Select the "junior-man" (the Japanese have a word for this) from the office. At this time of year, he or she will be a fresh college graduate with a new suit, shiny shoes, and a stunned, deer-in-the-headlights look about him or her. Dispatch this individual with a large tarpaulin to Ueno Koen (or another nearby park that stays open after dark) with orders to stake out a territory, and, under no circumstances, leave that guard post. If you're feeling generous, send a couple of friends to keep him or her company and to fetch water and sandwiches. It will be a long day of guard duty. Now, have your assistant source a karaoke machine—probably from the cupboard under the printer—and ensure that the batteries are fully charged. Send someone trustworthy to buy the beer. Show leadership by leaving the office before 6:30 p.m., and ask everyone to follow you to the park. Maybe let your hair down—loosen that tie a little; heck, maybe even take it off. (They're snappy dressers in Tokyo.) There will be a fun evening ahead for everyone— the Japanese really know how to party and enjoy themselves when you put a microphone in their hands. Lastly, as a manager, you would be responsible to wind things up in time to ensure that everyone can catch their train home.

So, here is the problem with *Hanami*. It's not like the Emperor's Birthday— you can't just schedule it on your calendar. The cherry trees bloom when it suits them. This depends on the winter weather conditions and the arrival of spring. Yes, there is variation! The thing is—to really enjoy the spirit of it all—you have to hold the party under the trees while the cherry blossoms are falling around you—not before and not after. The whole idea is to celebrate the fragility and the shortness of life and its beauties. The window of opportunity is only a few days long each year.

So, what if you had to plan this party? What if you'd like to invite some visitors from the New York office for a bit of team building? When would you advise them to come to Tokyo?

For the Anderson family, the challenge is to visit Japan in order to enjoy *Hanami* with relatives. That's me pictured in the Asakusa Temple gardens last week with the junior carrier of the Agile Management gene code. When should we travel? We have to make a guess—airline schedules are more or less deterministic. So, we book for the first week in April, and then we set our expectations

such that we will probably get to see some cherry blossoms either blooming or falling, but probably not both, and maybe neither. Why the first week of April? Well, it is statistically the most likely period when the cherry trees will be in bloom and the weather ideal for *Hanami*. Mid- to late-March might be too early, especially if the winter was a cold one. Mid-April might already be too late. The cherry blossoms are around for only a couple of weeks, and the ideal conditions for *Hanami* last perhaps four or five days.

So, to run a successful *Hanami* party, you need to be prepared. You need to show agility and react to opportunity when it arrives. And for those hoping to enjoy the party on a visit to the Tokyo office, you need to set expectations accordingly.

In our real, home lives, we know how to deal with variation. We know how to set expectations appropriately. We accommodate variation as if it were natural, because it is. Why do we struggle to do so in our professional lives?

In conclusion, did we get to see the cherry blossoms on our family visit to Japan? Yes, we did! Did we see them falling? No, we didn't. Did we get to party under the stars, petals falling around us. Nope! Was our vacation satisfactory? Yes, it was! Because our expectations and our definition of satisfaction had variability built in.

Approximately 44 Minutes

Sunday, June 5, 2005

• •

Sylvia knew how to subtlely set expectations. She understood variation. By offering us approximately 44 minutes, she was setting our expectations . . .

• •

"OUR FLIGHT TIME, FROM TAKEOFF UNTIL LANDING, THIS EVENING WILL BE approximately forty-four minutes."

I was sitting in a small, single-aisled regional jet. Outside it was marked with the insignia of Delta Connection, a loose affiliation of aircraft operators that provide the feeder services to Delta Airlines. We were on the tarmac at Cincinnati, Ohio, bound for Pittsburgh, Pennsylvania. Our flight attendant was Sylvia. She was about 35 years old, with a Grace Jones–meets–Dennis Rodman hairstyle of short, crisply cut blonde Afro curls. It looked pretty good on her—in a sexy but no-nonsense kind of a way. It was her job to ensure our safety for the next three-quarters of an hour, while stuffing us with peanuts and ginger ale. It was dark outside the windows already. The 20 passengers sprinkled around the cabin were all keen to get to Pittsburgh and find their way to their respective beds, whether at home or in some homogenous business hotel—like the one I was destined for.

So, what is my point? The flight was going to take three-quarters of an hour, or according to Sylvia, "approximately forty four minutes"!

Now, hang on, we humanoids have five fingers on each hand. We tend to approximate in fives or tens or in multiples or halves. Three-quarters is a good approximation. We don't approximate in multiples of 11, like 44. So, what gives?

Sylvia knew how to subtly set expectations. She understood variation. By offering us approximately 44 minutes, she was setting our expectations that the flight time was scheduled as 44 minutes, but that it might vary within some range, say, 40 to 50 minutes. That would be normal, and we wouldn't be "late." In fact, the total scheduled time from push-back to docking was longer, and indeed, we arrived early in Pittsburgh even though the flight took 45 minutes.

So what is the lesson here?

Note the subtlety in language. It's variation-aware language. If she'd said, "Our scheduled flight time tonight is 44 minutes," we might have reasonably felt aggrieved when we landed after 45 minutes. Approximately 44 minutes is the language of variation. It's the language of a Deming-style agreement that conforms to a mid-point and spread. "Scheduled for 44 minutes" is conformance-to-plan and specification language. It sets the expectation that exceeding 44 minutes is out of specification, and it represents non-conformant quality. "Scheduled for 44 minutes" makes for unhappy customers. "Approximately 44 minutes" sets the appropriate expectation, and it makes for happy customers.

Prioritizing Requirements

Tuesday, September 6, 2005

• •

An enumeration of [Table Stakes, Differentiator, Spoiler] allows us to align strategic positioning directly with our marketing product-mix selection choices and with work done on the shop floor, actually coding the features required.

• •

THERE ARE MANY SCHEMES FOR PRIORITIZING REQUIREMENTS. THEY OFTEN take the form of enumerated sets, such as [1, 2, 3], or [High, Medium, Low], or [Must Have, Preferred, Nice to Have], or the one we're promoting with MSF, which started with the DSDM community in the UK—MoSCoW [Must Have, Should Have, Could Have and Won't Have]. So, here is the scoop—I don't like any of these!

Why?

Because they all need explanation!

They all need some mapping. They require some standard definition of what each term means, and how to interpret them. This is hardly ever written down; often it is part of the tacit knowledge, or tribal lore, of an organization. It doesn't lend itself to repeatability, and it doesn't lend itself to good governance. I want a system that is better aligned with good corporate governance.

An idea I've been batting around for a while, but haven't blogged, is based on strategic-planning language. The idea is to categorize requirements using an enumerated set consisting of [Table Stakes, Differentiator, Spoiler].

Michael Porter has suggested that, from a strategic-positioning perspective, your business can be the cost leader (there can be only one), a niche player (with a market share that is typically less than ten percent in an established market), or differentiated (there can be many forms of differentiation). I like to use the example from my recent past in the wireless telecom business in the USA. There were six major players: AT&T Wireless, Verizon Wireless, Cingular, Sprint PCS, Nextel, and Voicestream (later T-Mobile USA). They were aligned thus: T-Mobile was the cost leader—note, this position was inherited from Voicestream, a firm T-Mobile acquired—Sprint was the innovator, Verizon Wireless had the best network, and Cingular was something called "the value leader," while Nextel was a niche player in small- and medium-sized businesses with mobile workforces, such as HVAC supply and maintenance. That left AT&T Wireless, which occupied the space that Porter calls "stuck in the middle." Being in the middle is not good. Guess which got acquired first?[1]

So, what does this mean and how do you play it? Well, imagine you are offering a wireless email service. Your table stakes are all the basic features you need to be in that business. You need a network. You need handsets that are email-capable. You need data service. You need an email server, and so forth. Now imagine you are the cost leader—stop right there. You've already spent as much as you need to. But if you are an innovator, you might want to do picture email, in which case, you need to upgrade everything. If you are niche player in small business, you might want to offer Microsoft Exchange enablement and rich-client concepts like the ability to read .DOC files. If you are the best network, you might want a lot of non-functional requirements that ensure the best quality of service. All of these are the differentiating features for your product-mix selection. Finally, what is a spoiler? A spoiler is a feature that copies one of your opponents' differentiators. For example, when Sprint and Verizon Wireless launched "push to talk" walkie-talkie features on their services, they were spoiling a lucrative market occupied by Nextel.

An enumeration of [Table Stakes, Differentiator, Spoiler] allows us to align strategic positioning directly with our marketing product-mix selection choices and connect it to the work done on the shop floor, actually coding the features required - no need for any translation scheme. No need for tribal lore to obfuscate the risks being managed. Strategic planners can appropriately position the business against the forces of competition.; Marketers can then choose a product mix that correctly selects features to target the strategic positioning and the desired goals in a given market, for example, profits, share, or prestige.; From this information the software-development team and the project manager are able to correctly select the release and iteration backlog to deliver the product mix in a fashion that aligns with those goals and the strategic position of the business.

One big, happy family! And everyone lived happily ever after and they all retired rich from their company-matching stock in their 401K (tax-deferred retirement savings) plans. Awe!

---

1. Readers from the United States will be aware that AT&T Wireless is still a going concern. However, the firm using this name is, in fact, Cingular Wireless, from Atlanta, which acquired AT&T Wireless (the former McCaw Wireless) of Redmond, Washington, and later adopted its name.

Revisiting Prioritizing Requirements

Tuesday, December 26, 2006

· ·

In an IT department, it is possible to deploy a system
because the features delivered allow for business
reorganization or business-process changes that
reduce costs and increase efficiency.

· ·

SINCE ARRIVING AT CORBIS AND ASSUMING A POSITION LEADING SOFTWARE
development in an organization that doesn't directly offer a technology product
or service, I have come to realize that my three-point categorization scheme for
classifying requirements, [Commodity, Differentiator, Spoiler], doesn't cover all
the required options. There is a fourth category—Cost Reduction.

In an IT department, it is possible to deploy a system because the features
delivered allow for business reorganization, or business-process changes, that
reduce costs and increase efficiency. By doing so, these features effectively increase the profit margin on whatever product or service the business is selling.

Hence, cost-reduction features need to be prioritized* against the imagined
cost saving, or improved margin, that will be generated through implementation. They need to be mixed in and compared against differentiators and spoilers
when making a feature-mix selection for an iteration or a project plan.

· · · · · · · · · · · · · · · · · ·

*During this period, 2005–2007, I was still referring to this classification scheme
as a means of prioritization. I later realized that I had developed a classification scheme for assessing the market risk, or uncertainty, in a particular feature
or project. Risks should not be prioritized; rather, they should be allocated. By
classifying requirements into these four categories, I was enabling an allocation
of scope and the ability to manage and align risk with the nature of the business.
See the later post, Planning for Market Risk, from 2008.

Why We Are Not Ready for Real Options

Tuesday, February 26, 2007

· ·

We need to encourage software engineers to think
through decisions with questions, such as, "Are we
going to need it or not, and if we are, what would we
spend now to balance that risk?"

· ·

I ATTENDED A REALLY INSTRUCTIVE SESSION AT OOP 2007. HAKAN ERDOGMUS,
of the National Research Council Canada, presented "Principles of Software
Process and Project Decisions." In this paper, Erdogmus proposes that we adopt
Real Option Analysis to inform and frame decisions in software engineering and
in Agile project management. This isn't the first time real options has come up
for discussion. Chris Matts has been a proponent, and real options was an active
topic of hallway discussions at Agile 2006.

Real option analysis is a mechanism for making decisions that would help
us get beyond blunt management principles, such as YAGNI (You Aren't Going
to Need It). By its very definition, YAGNI presupposes that options are never
worth buying. The Agile guidance of YAGNI is a reaction to the assumption,
in traditional software-engineering advice, that an option is always worth buying. In traditional software engineering, Boehm's research on the cost-of-change
curve suggested that options to build high quality into our work early in the
lifecycle were worth purchasing—that pursuing very precise, high-quality requirements would payoff asymmetrically due to the exponential growth in the
cost of late changes. Traditional software engineering has generally encouraged
design for reusability. This implies that buying the option of reusability has a
positive payoff later.

In his book, *Extreme Programming—Embrace Change* (Addison-Wesley,
2e, 2004), Kent Beck proposed a different cost-of-change curve that shows that
buying an option based on quality early in the lifecycle is not a good bargain
because the cost of fixing a problem later is actually much lower than the traditional curve suggests. In reality, we cannot generalize about the cost of change—
both curves are wrong. As I pointed out in my book, *Agile Management for
Software Engineering: Applying the Theory of Constraints for Business Results*,
the cost of change depends on the position of the constraint (or bottleneck) in

the software-engineering workflow, and in the variability inherent both in the domain and around the methods and skill level of the practitioners performing the work.

Real option theory suggests a solution to this problem. Real options offers a framework for decision making that will work in specific contexts, rather than encouraging the use of general, context-free guidance. Advice such as YAGNI is blind to the true cost of change in a specific project. General rules should be avoided. Instead, it would be better to teach knowledge workers a framework for making context-specific decisions.

So, real option–based decision making is desirable! However, I believe that we aren't ready for it—neither as an industry nor as a profession. Following are the reasons why I believe that.

Real option theory requires us to calculate two different numbers with some degree of confidence. The first is cost. We need to be able to calculate the cost of "buying" an option. We need to be able to accurately apportion the effort involved in, say, investing in higher quality early in the lifecycle, or in designing a class to be reusable, as opposed to foregoing reusability and minimizing the design of the class. This effort estimate must then be turned into a cost estimate. This amount becomes the cost of "buying" the option. Next, we must be able to predict the variation in—or the probability that we would take up—an option, or that circumstances would develop such that we would wish to "exercise" the option; that is, avoid spending time fixing bugs late in the lifecycle, or reuse a potentially reusable class or framework on a future project or iteration. Once we have both of these pieces of information, we can make an informed option-theory decision—either to buy or to pass on an option—based on whether the cost of the option is less than the risk-adjusted potential loss from not buying the option and suffering the consequences later.

The reality is, as an industry and a profession, we are years away from having the maturity to correctly measure and assess this kind of data; hence, I can only conclude that the day-to-day use of real option–based decision making is still a long way off for software engineering. Where I feel we can salvage something from this is the paradigm of option-based decision making. We need to encourage software engineers to think through decisions with questions, such as, "Are we going to need it or not, and if we are, what would we spend now to balance that risk?" Getting software engineers to think about early-lifecycle decisions as "options" would be a step forward to delivering better project decisions that are tuned to the specific situations and organizations, rather than decisions based on generalized assumptions about the cost of change or the likelihood of reuse, even if those decisions are not based on reliably informed, objective data.

Exploring Real Options–Based Software Engineering

Tuesday, February 27, 2007

● ●

In Lean terms, description definitions are postponed until runtime. Building a (blue) description class using a table-driven or XML file–driven approach is buying an option to postpone description definition until runtime.

● ●

I'D LIKE TO FOLLOW UP MY POST ON WHY WE AREN'T READY FOR REAL OPTION analysis–based decision making in software engineering with an example, which is based on Stephen Palmer's excellent article on modeling a library management application, originally posted on *The Coad Letter*. If we look at the model in Figure 14.2 (based on Peter Coad's Domain Neutral Component [DNC] pattern), we can see that it decouples and encapsulates many concepts.

In his article, Stephen Palmer explores the requirements space; he gradually narrows the possibilities and reduces the options in the scope. As he does this, he removes classes from the domain model. This demonstrates the "modeling by taking away" approach of *The Coad Method* in its final, 1999 version,[1] using color archetypes and the DNC pattern. There are many options; for example, the AccountInApplication class decouples MembershipAccount from its role in a Registration. Do we need this class? What does it buy us? If we create this class, what "option" is it buying us? We could say the same for the other (yellow) role classes, Library and Member. If we create role classes for Parties, Places, or Things, what "option" does that buy us?

Role classes decouple behavior associated with the transactional (pink) Moments, or Intervals of time, from the (green) Party, Place, or Thing (PPT) classes. With the inclusion of roles, the model is more loosely coupled and more highly cohesive. The (green) PPT classes are more reusable, and their responsibilities are more cohesive. If the (green) PPT class might be involved in another

---

1. Peter Coad did not develop his method further after 1999. He focused on the development of his business, Togethersoft, which was subsequently sold to Borland. He is retired from the field of software engineering.

transactional (pink) Moment-Interval, or even a whole other application, then the (green) PPT class is not polluted with behavior associated with several applications or transactions. This makes the classes cleaner, simpler, of higher internal quality, and probably easier to test, which results in better external quality. But there is an even more useful purpose to separating out (yellow) role classes from (green) PPT classes. It allows postponement of component boundaries and separation of code into discreet, coarse-grained components, packages, or applications, as explained in the next article, Coarse-Grained Components from a Color Model.

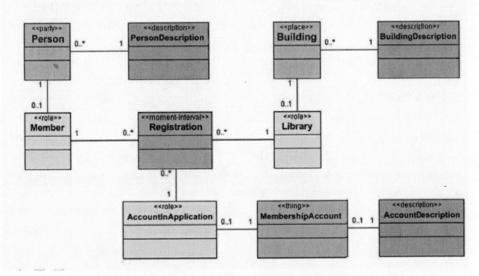

Fig 14.1   Model of a Library Management Application (courtesy of Stephen R. Palmer)

The real option–theory question becomes a tradeoff between the cost of creating the (yellow) role class—the price of the option—and the likelihood that a (green) PPT will be involved in another (pink) Moment-Interval, and that there will be a desire to separate out the different transactional application pieces from the reusable (green) PPT and (blue) Description classes, and the desirability, or otherwise, of avoiding heavy refactoring on the green class. The likelihood will depend on the domain, the specific requirements, and the direction of the business in the future. As described in the next article, assessing the cost of building the extra (yellow) role class is almost impossible, and assessing the probability of the need to decouple transactional behavior and/or to partition it into discreet components is equally hard. Hence, the input data for the real option equation would be highly suspect at best.

Let us analyze another example, using the same model from Figure 14.2. The (blue) Account Description class allows us to separate out metadata that describes types of accounts and associated behavior—for example, the maximum number of books that can be borrowed by any member registered with this type of account. Description classes are often implemented as database tables, or as XML files that are loaded dynamically. Both approaches allow for the metadata to be updated at runtime. In Lean terms, description definitions are postponed until runtime. Building a (blue) Description class using a table-driven or XML file–driven approach is buying an option to postpone description definition until runtime. To assess this option, we must again be able to account for the cost of building the table access or the dynamic XML file–loading infrastructure. We must also be able to make an assessment of the likelihood that the description definitions will change, how often this is likely to happen, and with how much notice. Is it likely to happen so often and with sufficiently short notice that deploying a new version of the software would be undesirable? Would we prefer to have a system administrator or a "super user" update the definitions, rather than involve the programming and testing team? If we can assess both the cost of buying the option, and the likelihood of the option being exercised, we can use real option analysis to make a decision about the value of building the (blue) Description class, and whether it would be better to build a table-driven implementation or an XML file–driven implementation.

Is your organization capable of making these kinds of well-informed, real option–based architectural and design decisions?

Planning for Market Risk

Wednesday, September 3, 2008

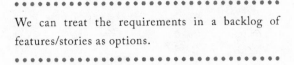

We can treat the requirements in a backlog of features/stories as options.

THREE YEARS AGO THIS WEEK, I INTRODUCED A SCHEME FOR PRIORITIZING requirements that was aligned with strategic planning and market segmentation, and it brought some rigor and objectivity to the art of assigning priorities to requirements.

Initially, the scheme introduced three classifications: Commodity (or Table Stakes), Differentiator, and Spoiler. Later, I realized that a fourth category, Cost Saver, was required. With two possible sub-classifications of cost saved by the customer (lower cost of operation or displacement of other costs), or cost saved to you, the producer of the software or operator of a service. These four classifications have survived almost two years without revision, and they seem to be quite stable. I think of this system as a lightweight, and very fast, alternative to Kano modeling.

Since 2006, I've been exposed to Chris Matts and Olav Maassen's Real Option Theory thinking. Some of you might recall that I was skeptical about real options after seeing a presentation in Munich, in early 2007, that attempted to use the Black-Scholes equation from financial options theory to assist in prioritization decisions. I pointed out that we, as an industry, weren't ready for real options because we couldn't furnish the equation with sufficiently accurate data. Well, the Matts-Maassen approach to real options does without Black-Scholes, and it is the better for it. Matts-Maassen tells us to push back decisions as far as possible, and to gather information, create options, and understand when they expire. This helps us to optimize decision making and minimize the risk of making a bad decision.

We can treat the requirements in a backlog of features or stories as options. Then we can consider the market risk and the likelihood of change occurring to a specific requirement based on its classification, as shown in Figure 14.2. We see that commodity/table stakes features have the lowest market risk. When did you ever hear of the table stakes going down? Differentiators have the highest market risk because, from a market perspective, they are the most uncertain—they represent untested assumptions.

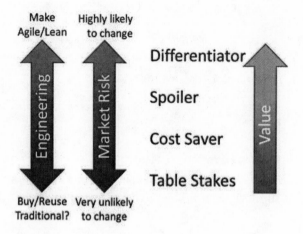

Figure 14.2 Mapping requirement classifications to market risk

Table stakes features are unlikely to change because the table stakes for market entry never go down. They only ever go up.

Cost savers are more likely than table stakes to suffer change during the life of a project, but it is still very unlikely. There is some risk that the cost is mitigated in other ways, or that shifts in the market, the strategic plan, or the go-to-market initiatives of the business might deprioritize the need to save costs. Hence, there is some risk that cost savers might be dropped from scope or be unneeded before the project is delivered.

Spoilers are features introduced to spoil a competitor's differentiator. They primarily protect market share. The main risk with spoilers is that by the time they are introduced, they have essentially become table stakes. This doesn't really affect our planning decisions. A planning decision would be affected, for example, by a choice to focus on a different market, or to concede an existing market to a competitor with a more compelling offer.

Differentiators are the most risky because they might turn out to be features that the customer does not want. They are essentially experimental in nature. If the customer likes them, they are worth a lot of new profit margin; if not, they don't produce profits. There is also a chance that a delayed differentiator is, in fact, a spoiler by the time it comes to market.

Now, applying what we know about market risk, and using a real-options approach to decision making and scheduling, we should delay risky items that are likely to suffer change until the last possible moment. This implies that we should code all of our commodity features first, and we should delay the differentiators until we're closer to the release date, as shown in Figure 14.3.

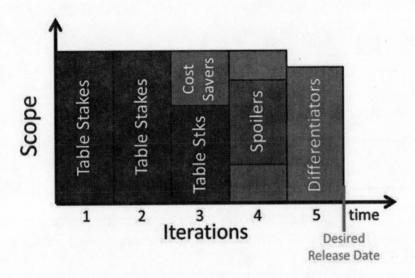

Figure 14.3 Planning a series iterations using market risk and feature classification

• • • • • • • • • • • • • • • • • • •

This approach is presented in a somewhat simplistic fashion. It suggests that we are addressing only a single market niche and that there is no likelihood that early or partial delivery would be required. In reality, some mix of table stakes, cost reducers, spoilers, and differentiators is required. It is, however, reasonable to suggest that most table stakes are scheduled early, and that many differentiators should be deferred until closer to delivery.

A Real-World Example of
Analyzing Market Risk

Wednesday, September 3, 2008

● ●

If you are able to deploy a differentiator prior to completion of the all the table stakes features, you simply haven't done your market segmentation and feature classification properly.

● ●

A POST IN THE AGILE MANAGEMENT YAHOO! GROUP HAS PROMPTED ME TO make this second post, with an example, to elaborate on prioritizing and planning for market risk.

Steven Gordon wrote:

> This approach seems oversimplified to me. In particular, I challenge the assumption that the low risk items should always be implemented first. For example:
>
> 1. A company might not even choose to bother with the market unless they can bank on a successful differentiator. In this case, shouldn't at least a slice or two of the intended differentiator be developed very early to obtain market research feedback on whether the idea really is a differentiator, whether it is really feasible to develop, and whether it might need tweaking or rejection.
>
> 2. A truly innovative differentiator can reshape the market and development vision of the whole product. Such a paradigm shift could totally change how the commodity features should be presented and developed.
>
> I am more comfortable with the general notion that features should be delivered in order of immediate value than risk or risk of change. So, differentiators that have a large impact on how the whole product would be designed or on whether the project is even worth doing have large immediate value.

Here is my reply:

I believe a real-world example would help. Let's imagine that we are going to enter the cell phone market. Let's first define the table stakes/commodity features. Because entering the cell phone market has a huge setup cost, it will be necessary to offer a broad service that is attractive to many market segments. As mentioned in the previous article, what I presented there deals only with a single market segment. For the purposes of the example, let's try to define the general subset of features that represent table stakes for entering the cell phone market, treating it as a single segment.

We will need the following:

- ❐ A cell tower network
- ❐ A deal with a long-distance carrier to back haul our calls
- ❐ A PSTN/SS7 telephony stack and associated infrastructure
- ❐ Dial-tone
- ❐ Ring-tone
- ❐ Call setup across multiple networks
- ❐ Call teardown across multiple networks
- ❐ A billing system with the ability to bill individual calls and reconcile billing across networks to terminating or originating networks
- ❐ An SMS gateway
- ❐ A voicemail system
- ❐ Handsets (one model might suffice, but that is unlikely) with firmware to support voice calling and SMS messaging
- ❐ Address book feature on the handset
- ❐ Inbox feature on the handset
- ❐ A support infrastructure and call center
- ❐ A channel to market, such as a retail network

Although we could argue the finer points of this set of features, the evidence from the market suggests that they are broadly correct. For example, if we study the mobile virtual network operator (MVNO) market entrants in this century, such as Virgin Mobile USA, we can gain insights to the true table stakes. Virgin rents its network from Sprint. At the time the deal was struck, Sprint did not offer SMS messaging. It offered a rival service and technology from Qualcomm. Virgin thought it sufficiently important to include SMS that they deployed their own gateway to augment the Sprint network infrastructure. Virgin also sought to target a market niche—teenagers. To enable this, they required prepaid billing. Sprint did not offer prepaid billing at that time; later, they set up a separate brand for prepaid service. Virgin, again, acquired their own prepaid billing system to enable entry to the market for teenage cell phone users.

Whereas other features might be argued as "table stakes," the reality is that they are probably table stakes for specific market segments, rather than for the general market. This is a common problem with many firms. They do not perform good market segmentation, and they fail to truly understand which features of their product or service relate to specific segments. As a result, they believe the table stakes are higher than they are. Smart market insurgents can exploit this situation by offering bare-bones, or white label, products or services that offer just barely sufficient functionality to a market niche that needs unsophisticated features. A recent example of this is the market entry of Google Docs, which started life as a barely sufficient word processor that was only slightly more advanced than the Notepad application shipped as part of the Windows operating system.

Steve Gordon suggested that differentiating features should be developed early in order to test them in the market and prove their value and market acceptance. I believe this is confusing research with development of production systems. I believe these are parallel activities. Returning to our cell phone example, let's imagine that a differentiator is picture messaging (MMS). It would not be possible to demonstrate this feature on the production network without first delivering the table stakes. Although it is technically possible to skip the deployment of the SMS gateway, there is an issue with market acceptance if the SMS service is not offered.

However, the cell phone operator could test the market for picture messaging using research equipment. A cell on wheels (COW) could be deployed for market research. Typically, COWs carry new network technology that is not deployed in production. COWs demonstrating full Wideband Code Division Multiple Access (W-CDMA) technology were available in 2001–2002, although most operators have yet to deploy W-CDMA for general use at the time of this writing, in 2008.

If you are able to deploy a differentiator prior to completing all the table stakes features, you simply haven't done your market segmentation and feature classification properly. The table stakes that are not yet completed are probably related to specific segments, and not to the segment where you are testing the new functionality.

Steve states that value is more important than market risk or risk of change. Really? As recently as three years ago, I probably agreed with him, but I've come to realize that I was wrong. Why?

First, value is very difficult to define. Value is contextual, and context is temporal. At best, definitions of value are crystal ball gazing, and they are projecting an estimate based on assumptions and on today's market conditions. Let's imagine that a new feature envisaged as a differentiator is planned. Let's say it is picture messaging, and market research suggests that it is worth a $10 per month increase in the subscription fees. The lead time to develop it and deploy it is 18 months. A year into the project, a rival network launches picture messaging at $7 per month. What is picture messaging now worth? First, it is no longer a differentiator; it is a spoiler. And it is probably not worth even $7 per month, but somewhat less. Either the price falls, or the size of our target market falls, because the insurgent stole some of our market share while we waited to bring the feature to market.

Let's now imagine that the competitor launches the feature early in our development cycle, say, month two. The feature is a differentiator for a small market niche. If we follow Steve's strategy of prioritizing value first, we are building this differentiator first. We have started work; we have sunk cost into it. When the competitor launches, we redo our market forecasting and come to the conclusion that the feature is no longer viable. The cost outweighs the perceived gain in the market, so we cancel the feature and order another one. We just wasted valuable energy that could have been used developing a lower-risk table stakes, cost saver, or spoiler feature.

Hence, there are two reasons for not pursuing a value-first strategy. The first is that value is hard to define and there is a high variability to it. The second is market risk that puts the viability of the feature at risk. It's simply bad use of real option theory to prioritize something before we have enough information to be sure that we want the feature delivered.

Banish "Priority" and "Prioritization"

Sunday, March 20, 2011

Eliminating proxy variables empowers team members
to make dynamic, good-quality risk decisions.

THOSE OF YOU WHO'VE ATTENDED MY CLASSES OR WORKSHOPS, OR WHO HAVE talked with me at conferences this past year, or who have read the Kanban book very carefully might notice that I have purged the use of the words "priority" and "prioritization." I'd like to explain why I've done that.

"Priority" is something that Don Reinertsen would refer to as a "proxy variable." It is an artifact that masks real risk information, such as "cost of delay," required skills, technical impact, transaction cost information, and so on.

Our goal with a Kanban system's visualization is to find ways to capture and visualize the true risk information that the business is managing. Classes of service do this very well, so long as the policies that define what it means for an item to be of that class—and how an item of that class should be handled—are made very explicit.

"Priority" then becomes something that can be decided dynamically when a pull decision is required.

Eliminating proxy variables empowers team members to make dynamic, good-quality risk decisions. It reduces the need for coordination meetings and it improves transparency. It also obviates the role of middlemen who determine "priority." This largely explains why a role such as Product Owner is generally not required with Kanban.*

The term "prioritization" is also no longer required with Kanban. Prioritization is implied by the policies associated with a set of classes of service, and how each class of service interacts with the others. For example, Expedite takes precedence over Fixed Date, which might take precedence over Standard, depending on its current position in the workflow and the likelihood of on-time delivery against the fixed date.

When we select something to replenish a slot in an input queue, I prefer the terms "selection," "scheduling," and "replenishment," rather than "prioritization," as we aren't prioritizing a set of things in a backlog, we are selecting something from a (usually) unprioritized backlog to place in our input queue.

Scheduling implies that we chose to select an item for queue replenishment based on some plan or delivery schedule, or based on another dimension of risk being managed, such as market risk of change. We might choose to schedule table stakes features early. This means that they will be given priority when we are replenishing the input queue early in the project. Scheduling is definitely an optional practice, whereas selection and replenishment are necessary to facilitate the flow in the kanban system.

When we select something to pull from an earlier stage in a workflow, again, I prefer the terms "selection" and "pull," based on the policies in the classes of service package and the pull criteria (also known as "definition of done," or "exit criteria").

So "priority" and "prioritization" go away. They are replaced with "risk profile" and "class of service" (to replace "priority"), and "selection," "scheduling," and "replenishment" (to replace "prioritization").

In Lean terms, I find that "priority" and "prioritization" are wasteful. They encourage roles/positions for people who do mostly non-value-added coordination work, and they add to the transaction costs of flowing work through the system. Prioritization is an activity performed at a point in time that presupposes to predict the future. Deferred decision making is a more Lean approach. Kanban encourages deferred decision making and dynamic prioritization based on policies written to accommodate the risks being managed.

I am finding that by encouraging teams to abandon the use of words such as "priority" and "prioritization," which are associated with an older paradigm, a mindset shift to a flow-based approach is easier to achieve. This mindset shift improves the quality of the kanban system design and risk management. Better risk management should lead to better satisfaction for all stakeholders.

● ● ● ● ● ● ● ● ● ● ● ● ● ● ● ●

* Where the role of Product Owner is observed in a Kanban implementation, it is generally an artifact of the process in use before Kanban was adopted. Under the core principles, "You start with what you do now," and "Initially, respect current roles, responsibilities, and job titles," if what you do now involves a product owner, that role is likely to continue for some time, until it is generally accepted that the role has been obviated, and a new role is found for the individual currently playing the product owner.

Wardrobe Triage

Wednesday, October 18, 2006

I'VE DECIDED TO TRIAGE MY WARDROBE! "HUH?" I HEAR YOU saying.

America is big. Things in America are big. Cars are big. Houses are big. Parking spaces are big. Burgers are big. Fries are bigger. Most of all, spaces for hanging clothes are big. In fact, even in a relatively densely populated coastal city like Seattle, walk-in closets tend to big enough for a dad to swing a small child. The problem with this is that it is easy to keep shelving more and more clothes and never throw any out.

It's also really easy, with approaching middle age and fast-growing children, to let years go by and forget about wardrobe updates. Then one morning comes when you look in the mirror and realize you are dressed in clothes that predate your marriage. So, I've decided to triage my wardrobe.

When I was on my Asian tour recently, I was surprised at how often I had to explain the triage process for software projects. Typically, triage is held to process and prioritize bugs reported in system testing. However, I suggest that triage is also a good process for sorting work queuing for development. In MSF for CMMI* Process Improvement I've provided a triage report that shows all the work-item types— such as requirement, change request, bug, issue, and risk— queuing for prioritization. It seems that the meaning of triage is not widely understood. So, I thought I'd use the personal fashion disaster I call a wardrobe as way to explain it.

Triage is a process borrowed from medicine—initially, field emergency medicine. It is the process used to sort patients in an emergency (A&E) room. The triage nurse examines patients and decides whether they: need immediate attention; need attention, but can wait until later; do not need attention (in a wartime field-hospital situation, this usually means that they are beyond help).

In software triage, we seek to sort things into three buckets, too: things that need to be dealt with in the current iteration; things that can be postponed to a future iteration; and things that will not be dealt with (that is, not fixed, not selected, rejected, and so forth).

In order to make the selection, we need a set of criteria that can be applied consistently to each item in the backlog. At Microsoft, these criteria are described as a "bug bar."

So, when it comes to triaging my wardrobe, I need a set of criteria with which to process the clothes hanging on the racks. So here is a first-pass set of criteria:

- ❒ I haven't worn the item for more than two years.

- ❒ The item is more than ten years old.

- ❒ The item is so out of fashion I can't imagine wearing it ever again.

- ❒ The item is faulty or has developed wear or holes.

- ❒ The item does not fit (I lost a lot of weight since I left Kansas City five years ago).

All items meeting at least one of these criteria will be placed in the third bucket—the one headed for a Salvation Army store.

- ❒ The item is out of season but not out of fashion.

Summer clothes aren't needed now that fall (autumn) is with us. However, they can be "postponed" for use next summer. I'll sort these to the back. These represent the second bucket—clothes postponed for future wear.

That will leave me with the current fall/winter clothing. I can then do a gap analysis (against my unwritten requirements) on these and determine what I could do with buying to complete my new fall/winter wardrobe.

On Roles, Responsibilities, and Organizational Structure

MAYBE YOU THINK MANAGERS ARE MERELY "CHICKENS," WHO ARE INVOLVED WITH, BUT NOT RESPONSIBLE FOR, "REAL WORK." MAYBE YOU THINK MANAGEMENT IS SETTING THE POLICIES THAT WILL MAKE OR BREAK YOUR LEAN OR AGILE INITIATIVE. IN EITHER CASE, SHOULDN'T WE CONSIDER the plight of management? Ultimately, management is held accountable for the success or failure of an organization's efforts.

Kanban strives to optimize the flow of work through the development cycle. This is not a static thing. To assist, managers need to shift their roles and their policies continually to meet the current conditions. They can affect variability, bottlenecks, the input queue—so many of the things that enable flow to happen.

If you are a manager and you are held accountable, what would you want or need to do your job? You would want visibility into the work and the process. You would want to be involved in decisions as they are being made. You would want a collaborative culture in which people make promises that are real and that are based on capacity.

The Agile world has focused on the development team and the code, while the Lean world has pushed toward enterprise-wide cultural change. For true business agility, everyone must feel empowered, not just the development team, and organizations must be structured to perform large-scale, complex work in a coherent fashion. Some attention to systems architecture, and assigning organizational units to highly cohesive work through robust, loosely coupled interfaces is required.

ARCI not RACI

Monday, January 10, 2005

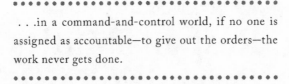

. . .in a command-and-control world, if no one is assigned as accountable—to give out the orders—the work never gets done.

RACI IS AN ACRONYM USED IN FORMAL SOFTWARE-ENGINEERING PROCESSES TO designate roles and responsibilities. The letters mean:

R: Responsible (does the work)
A: Accountable (blamed when the work is not completed satisfactorily)
C: Consults (provides input)
I: Informed (receives the output, or a facsimile or summary thereof)

The problem with RACI is that it sounds nice, but it warps the focus. It ought to be ARCI—perhaps not so English-language friendly, but it more accurately puts "accountability" first on the list. Too often I see process specifications in which the author (in a hurry) omits the A and specifies only who is responsible to do the work—fine, you might think. Well, the problem with this is that in a command-and-control world, if no one is assigned as accountable—to give out the orders— the work never gets done.

ARCI is firmly a facet of the command-and control-management style, and it belongs in the traditional software-engineering world. How is the Agile community adapting this concept for a higher trust, more empowered culture?

Chickens and Pigs

Tuesday, January 11, 2005

• •

This CI, or pig and chicken designation, removes the problem of command and control by merging the doing with the giving of orders.

• •

KEN SCHWABER INTRODUCED THE CONCEPT OF CHICKENS AND PIGS WITH THE daily scrum. Each attendee is either a chicken or a pig. The metaphor comes from the idea of a project to make a breakfast of bacon and eggs. In such a project, the pig is said to be committed, but the chicken is merely involved.

Committed and involved could be mapped to the ARCI designations for roles and responsibilities. This might help to map an agile process into a more formal, or traditional, organization.

Committed (pig) is the equivalent of A(ccountable) and R(esponsible).

Involved (chicken) is the equivalent of C(onsults) and I(nformed).

This CI or pig-and-chicken designation removes the problem of command and control by merging the doing with the giving of orders. A committed designation allows for empowerment, and its alter ego, self-organization. Committed allows for shared responsibility (in the common sense) and shared ownership. Everyone on the team who is doing the work is committed, and they are all jointly accountable.

Chiglets

Wednesday, January 12, 2005

• •

When everything is running well, the line manager
is pretty much superfluous. However, under chaotic
conditions, the line manager is a useful member of
the team.

• •

LINE MANAGERS COMPLAIN TO ME THAT THEY ARE NEVER SURE WHETHER THEY
are chickens or pigs. Should they even be in the morning standup meeting? I've
vacillated on this question in the past. My opinion, now, is that when everything
is running well, the line manager is pretty much superfluous, unless he or she
also plays another role, such as architect or domain expert. However, under
chaotic conditions, the line manager is a useful member of the team. So, is a line
manager a chicken or a pig?

Line managers are seldom scrummasters. Scrummasters usually report to
line managers, in my experience. So, in theory, the line manager is a chicken—
merely involved; someone who receives notification and output, and who occasionally is consulted for input. However, when an assignable-cause event
happens and an escalation is needed in order to remove an impediment and get
work moving again, suddenly the line manager is a pig—committed, doing the
work—both accountable and responsible.

Hence, line managers live in a world that oscillates between two states, never
quite stable as either a chicken or a pig. They are chiglets! If we're not happy
with the idea that a chiglet isn't permanently committed, and sometimes merely
involved, I suggest that we designate the chiglet's role and responsibility relationship as "facilitates."

Hence, an Agile ARCI model would have three designations:

F—facilitates

C—committed

I—involved

Where Everyone's a Pig

Originally written for Borland, in 2009, prior to the acquisition

. . . .

When managers are excluded, their policies become impediments to adopting Agile techniques. Lack of inclusion, coupled with lack of transparency, breeds fear in managers, often causing them to react with inaction and immobility.

. . . .

I HAVE BEEN WRITING RECENTLY ABOUT HOW SUCCESSFUL AGILE transformations hinge upon true cultural change.[1] One of the themes that I've touched on is that a new Agile process cannot simply be mandated. Teams need to understand the reasons for the change, and they need to internalize the ideals and concepts of the Agile movement. They have to think with an Agile mindset to truly live it.

In the Scrum methodology, a clear distinction is made between team members and "the rest of the world"—which includes management and anyone from either an upstream activity, such as marketing, or a downstream activity, such as IT operations and systems support. Team members are referred to as "pigs"; the others are "chickens." The metaphor comes from the story of the chicken and the pig discussing a full breakfast that includes ham and eggs. The pig points out to the chicken that while she is merely involved, he on the other hand, is totally committed.

The Scrum framework implies that team members—those building the software and testing it—are the only ones who are committed and accountable. Meanwhile, managers and value-chain partners (stakeholders from upstream and downstream functions) are treated as outsiders. They might be involved—someone on the team might need their input, insight, or assistance—but they are not committed.

I've come to believe that this division is a direct impediment to a successful, enterprise-scale Agile transformation. Here's why:

---

1. http://www.borland.com/agile-transformation-forum/creating-an-agile-culture.html

Negotiation versus Collaboration

The rules of Scrum are all about making a development team much more productive. But scaling Agile means going beyond the development team to make an entire organization more productive and efficient. Succeeding at enterprise-scale Agile is all about creating an organization-wide Agile culture.

Part of the reason Agile has been such a success for development is that it gives individuals and teams ownership, and it empowers them to make their own decisions. For true business agility, everyone must feel empowered, not just the development team. Governing policies should be agreed upon collaboratively. To extend a culture of Agility across the value chain, you must educate upstream and downstream stakeholders, help them to internalize the Agile ideals, and then let them take ownership. You must get everyone committed to delivering the business goals. You must make everyone a pig.

In 2007, I wrote an article for the *Cutter IT Journal*1 that described how to change the culture in an organization and raise the level of trust (or social capital) by eliminating negotiation and creating cross-functional and cross-hierarchy collaboration through the use of cooperative, collaborative games. I had discovered that the psychology and sociology of inter-departmental interaction within the enterprise was greatly improved when individuals were asked to collaborate on solving a puzzle related to real work.

For example, when development teams using the Kanban Method interacted with their upstream colleagues to prioritize new work into their input queue, they asked the business and marketing folks questions, such as, "This week we have two slots free in the queue. Our average cycle time is 30 days, and our due-date performance against that time is 70 percent. Please pick the two things from the backlog that you most want delivered approximately 30 days from now."

What happened was remarkable. First, the group developed a democratic voting practice to resolve the challenge of selecting just a few items from a backlog of perhaps hundreds of ideas. Later, they devised an even more sophisticated collaborative approach to this. There was no negotiating. No arguing over whether the queue could be increased, or whether three small things could be squeezed into the two open slots. Negotiation had been banished. How did this happen?

Transparency and Inclusion in Policy Making

When the development team first decided to adopt the Kanban Method, they approached their value stream partners, including the upstream stakeholders who created the requirements—executives from sales, marketing, operations, finance, and human resources. The team explained the new process to these stakeholders, and together they discussed and agreed upon the new policies that it would introduce. Specifically, everyone achieved a common understanding of the team's capacity and a mutual agreement on the limits to work-in-process. Kanban's inherent pull system[1] mechanism was explained, and the policies around refilling the input queue were agreed upon. Because the business stakeholders could be held accountable for the deliverables (not necessarily for the actual development work, but, for example, sales is accountable to customers for features and functionality they have promised), in this scenario, everyone in the room was a pig.

Once the team started to use the new process, they made it transparent to all. They tracked their work on a white board, using sticky notes for each work item. These were also tracked using a software tool that produced electronic reports to share among the stakeholders.

The combination of giving the value stream stakeholders a say in setting policies that governed the collaborative game of software development and providing day-to-day transparency changed the relationship between the business and the software development team. Because the external stakeholders were educated on the process, they understood what impact their actions could have, and they didn't negotiate to push work on the team and break the process. Further, because they were included in the decision-making process, they felt ownership, and this naturally led to greater collaboration, a more effective process, and faster acceptance/adoption.

In another example, as Borland's development teams transitioned to Agile, they began to encounter challenges around this same issue: how to manage and prioritize very long backlogs that included requests from a wide range of sources. Because the business stakeholders knew that the Agile teams revisited the backlog before every sprint, ongoing negotiations and bartering began to take place. This was highly disruptive to the teams. Further, the "chickens"—both management and the value chain partners—felt like they couldn't really get a sense of the work that was going on in engineering, and this was having an impact on the perceived success of the transformation.

Borland's solution to this problem is similar to the earlier example. They established an Agile road-mapping process: Each quarter, the teams and the value chain partners (representatives from marketing, sales, and support) come

1. http://en.wikipedia.org/wiki/Kanban

together to discuss and agree upon the direction of the products and the prioritization of the queue. As in the previous example, implementing this involved educating the value chain partners and then giving them a voice in the process. So, in Borland's Agile road-mapping process, everyone is a pig.

Collaborate with Managers to Enable Agile Transition

In a similar vein, keeping management at arms' length, calling them chickens, and excluding them from talking at standup meetings is not helpful. Only through inclusion will management gain understanding of the impact their actions can have on the team and its performance.

Process is often described as a workflow—a series of assigned steps and artifact handoffs. However, you can also describe a process as a set of policies. In this case, the policies are essentially the rules of the collaborative game that we know as Agile Software Development. Ideally, in a highly empowered Agile team, most of these policies are under the control of lower-level managers and individual contributors. However, riskier decisions are likely to be governed by policies that require the authority of higher-level managers to change.

Most process improvements are achieved through changing policies. For example, a bottleneck caused by a specialist resource that is multi-tasking across several jobs or several teams can be resolved by changing the management policy relating to that individual's job description and responsibilities. Involving management and giving them the same transparent access to the process and the work flowing through it allows managers to see the effects of their decisions and of the policies that they control. They have the power to make changes and to influence the performance of the team. Further, management is the team's interface with the "outside world" and the complex set of factors that are involved in the value chain. If they feel ownership in the Agile process, they can help with the interactions and cross-functional policy making.

In the Agile organization, managers need to be pigs along with everyone else.

By bringing management into the process through a collaborative inclusionary approach, Agile transitions become easier. Managers can modify policies and push authority down to where it belongs. When managers are excluded, their policies become impediments to adopting Agile techniques. Lack of inclusion, coupled with lack of transparency, breeds fear in managers, often causing them to react with inaction and immobility.

Conclusion—There are No Chickens

Collaborating with upstream and downstream stakeholders as well as with managers gives everyone ownership in the team process. It breaks down the barriers and removes the "them" versus "us" divide. Bringing partners and managers into the process, providing them with visibility, and educating them on the cause and effect of policies, actions, behaviors, and activities allows them to adapt and modify their actions and policies, which enables faster Agile adoption and a smoother, more effective Agile transition. Make everyone in the organization a pig. Banish chickens altogether!

The Loosely-Coupled, Highly
Cohesive Business Unit

Tuesday May 9, 2006

• •

Encapsulation, the idea that an organizational unit
can work on stuff without outside interference is
amazingly liberating and powerful. We have a word
for it: Empowerment.

• •

ORGANIZING A BUSINESS FOR LARGE-SCALE AGILE ADOPTION REQUIRES SOME
thought. Here are some of mine.

What do I mean by larger scale? Typically, most Agile writing addresses only
single teams who are free to work on their own without interference, and who
single-task on a single project until it is completed. The main exception in the literature is Feature-Driven Development, as the method was created and designed
for use on medium- to large-size projects or programs (usually 50 people or more.)

Optimal large Agile organizations are likely to consist of small, highly cohesive teams that are loosely coupled.

Do you ever find yourself in one meeting after another with different sets
of people—and each time you are trying to agree on who will work on what
and how to avoid treading on each other's turf? Often, these meetings seem to
drag on for days, weeks, or months, and very little work gets done. Planning
large projects or programs often degenerates into a turf war in which managers
squabble about who owns what. A lack of cohesion across teams leads to a lack of
momentum because people don't know what to do or who to work with; or even
if they do know, the colleague is often on a different team and already assigned
to different work with a different priority. A lack of cohesion in the definition of
roles and responsibilities produces very large amounts of waste.

We might define this type of waste as the amount of time we spend talking about working as opposed to the amount of time we spend collaborating on
real work. A good ratio might be 1:10; a typical one might be 1:1; and in some
organizations, it might be more like 10:1 in favor of waste.

All of this leads to the main point—architecture is important in order to
scale Agile to large projects and programs. A well-defined, loosely coupled,

highly cohesive architecture enables an organizational structure that significantly reduces waste. It is precisely this idea that is enabled by using domain modeling in color to identify components and sub-systems in a robust fashion. Feature-driven Development takes advantage of this with its Feature Teams and the features packaged into a Chief Programmer Work Package. Because they are extracted from the domain model, the features within a work package are known to be highly cohesive. The Chief Programmer acts as the communication channel between groups of loosely coupled teams, each with their own Chief Programmer.

Although coupling and cohesion emerged as concepts in structured methods that predate object-oriented, aspect-oriented, and service-oriented architectures, the concept is still valid. The idea of a cohesive class, component, aspect, or service still holds. What is encapsulated inside the class, component, aspect, or service ought to be cohesive, and through its interface (its method signature), it ought to be loosely coupled. The same is true for organizations. Encapsulation, the idea that an organizational unit can work on stuff without outside interference is amazingly liberating and powerful. We have a word for it: Empowerment.

Architecture and analysis are vital to scaling Agile/Lean concepts across larger organizations. Without a loosely coupled, highly cohesive architecture on which to map the organizational structure, it is impossible to minimize the waste from communication overhead. You can't refactor organizations the way you can refactor code. Emotional intelligence is involved, and there is a lot of emerging evidence to suggest that the older, emotional part of our brains doesn't process and rewire as quickly as our logical processing centers. How often have you seen an employee complain about organizational change? As a member of my staff once told me, "You're the fifth manager I've had in 15 months and I can't stand it!"

At Microsoft, we live with these challenges every day. Around 450 people work on the Visual Studio Team System product. There are six product units with up to 150 people in each one—although some are significantly smaller. The Team Foundation Server product unit is split across two geographic locations, and organizing the groups in Redmond and Raleigh as cohesive, loosely coupled units has taken some time and effort on management's part. Much of the talking we do is about who owns what, how the architecture maps to the organizational structure, and—as our picture of the product and its function clarifies where each piece of work should land—determining who is responsible for the work and what the dependencies between teams are. Minimizing dependencies and coupling between teams and product units and creating clear lines of communication through loosely coupled interfaces—a job known as Program Manager at Microsoft—is part of daily life on a large-scale software project.

Adapting the Organizational
Structure to Change

Wednesday, May 10, 2006

· ·

Organization of businesses with people isn't as
easy as software architectures—we don't do aspect-
orientated organizations very well.

· ·

YESTERDAY, I TALKED ABOUT ARCHITECTURE, AND COUPLING, AND COHESION,
and object-orientation, and distributed components, and aspects, and services,
and how ideally we want a loosely coupled, highly cohesive organization
mapped against our product architecture. This provides an encapsulation that
is empowering to a team, which leads to high productivity with low overhead.
In Lean terms, our process has less waste if we eliminate all this coordination
overhead. However, organizing businesses, with people, isn't as easy as organizing
software architectures—we don't do aspect-orientated organizations very well.
Anything that has a cross-cutting concern, such as security, tends to get bogged
down in politics and negotiation over who owns it.

I'm not talking about matrixed organizations. We do that well already.
At Microsoft, we have Program Managers in charge of delivering features.
Individuals from function managers' teams are matrixed to the PM to develop
features in products. This works nicely. A feature team is, we hope, cohesive
and loosely coupled through the PM as the communication point. But it doesn't
work well when one feature is cross-cutting of others. This leads to a lot of extra
communication—waste—and slows things down.

There is a second problem in mapping architecture to an organization.
Architecture has uncertainties at the start of a project, just at the time when
you want to lock in the organizational structure. Furthermore, we tend to do
architecture top-down, and this usually means that we get it wrong the first time
and that several iterations are needed before it settles down. This leaves us with
an organization that is mismatched to the underlying architecture of the product—with the resultant overhead of extra communication, uncertainty, several
teams accessing the same code, and a general slowdown in productivity. While
developing Visual Studio Team System v1.0, we saw just these kinds of problems.

The MSF team was moved around several times, and parts of the process template and work-item tracking were also moved, as the general manager, Rick La Plante, looked for the best organizational structure to match the emerging architecture of the product. This puts strain on individuals, particularly product unit managers and the group managers who report to them.

However, there is a better way. Bottom-up architecture, driven off a domain model, has been shown to facilitate late definition of the component or service boundaries. This postponement of decision making fits with the Lean idea of making decisions at the last responsible moment. Bottom-up architecture solves the problems of organizing around a top-down architecture and the rework and confusion it can cause.

A few years ago, I worked on a project at another company. I was leading the team in Seattle. The architecture was being done in Mountain View, California, while the other development teams were in Beijing, Bangalore, and St. Petersburg. The architects analyzed the system and then divided the work among the four teams. What happened next? The teams spent months holding weekly global teleconferences to resolve issues with the interfaces between components and the roles and responsibilities between the teams. This led to poor productivity and lots of stress and confusion. This was a direct result of poor top-down architecture, and it was something that could have been solved by bottom-up, domain-driven architecture.

So here are the challenges I see still around organizational structure: How do you organize for the early lifecycle while you are building the domain model and deciding on the architectural separation of concerns? How do you allow for postponed, last-responsible-moment architectural decisions, and reconcile that with optimal, loosely coupled, highly cohesive organizational design? How do you organize for cross-cutting concerns? What does an aspect-oriented organization look like?*

● ● ● ● ● ● ● ● ● ● ● ● ● ● ● ● ● ●

*Editing this book and writing in 2012, some six years later, I am sad to report that I have yet to see a large-scale organization built for truly Agile and Lean software development. The questions I posed in 2006 remain unanswered. Perhaps it is time for a pendulum swing, back toward the greater structure and attention to modeling and architecture of the twentieth century?

Sushi Lunch

Sunday, April 10, 2005

SATURDAY, APRIL 9 WAS MY BIRTHDAY, ONE THAT PUT ME fairly and squarely in the late-thirties category (*san-ju-hatu-chi* to be precise). In honor of this occasion, my in-laws treated the whole family to a sushi lunch at a serious-but-not-overly-pricey sushi restaurant in the Hibiya district of Tokyo. Hibiya is a financial offices district to the northeast of the Imperial Palace. As it was *hanami* season and the palace gardens were busy with visitors, the restaurant, which was no more than a sweet five-iron over the moat from the palace grounds, was also busier than usual for a Saturday. However, this seasonal variation must be considered as common-cause rather than special-cause because it is entirely possible to anticipate that on any Saturday during *hanami*, the restaurant will be busier.

On arrival, we were told that we could not be seated—there was a constraint. The constraint was being protected by a capacity buffer of seats outside the door. We sat down. The menu was brought to us, but strangely, no one came back to take our order. Eventually we were seated inside at our table and a waiter arrived to take our order. My brother-in-law had put some serious effort into selecting the pieces to be ordered and the quantity of each piece, including some special treats for me, the birthday boy. The order was placed and we waited and waited. Meanwhile, the almost-three-year-old in the corner seat was getting hungry, and was eager with anticipation for the arrival of her California roll.

A plate of sushi arrived. No California roll!

On further inspection, it appeared that only half (*han-bun*) of our order had been fulfilled. An inquiry was made. A very polite waitress explained that the sushi chefs were fully loaded (a capacity-constrained resource), and that in order to keep all the guests happy and provide food in a reasonable lead time, they were filling only half the orders initially. The sushi chefs were self-organizing. They were presented with the order list and could burn it down as they saw fit. So, an essentially randomly chosen selection representing half of our order had been delivered. Fair enough! Small batch sizes. Incremental delivery. Shorter lead times. It all made sense.

However, the vocal toddler was annoyed and frustrated. She started to chant, in English thankfully, I WANT SUSHI! I WANT SUSHI!

This was a serious sushi restaurant. You could tell this because the only condiment on the table was soy sauce. The sushi was served with ginger, but no wasabi. This is to prevent uncouth foreigners from insulting the sushi chef by adding extra wasabi. The chef has expertly selected just the right amount of wasabi to enhance the flavor of each piece. Why add more? However, the uncouth do have an outlet, they can always dip pre-prepared pieces such as *unagi* (grilled eel, which already has soy sauce) into the soy sauce—*henna gaijin*!

So, we offered our unhappy customer the *tamago* (egg) sushi. She wrongly identified the egg as cheese and said, "I don't want cheese!" She dismantled the sushi, offered the "cheese" to Daddy, and ate the rice.

Eventually, the second plate of sushi arrived. No California roll!

This was a serious sushi restaurant! They could not, or would not, make us *Kariforunia maki*! A lot of bowing came with this explanation. The toddler burst into tears and wept uncontrollably, to the consternation of the waitress, who could not comprehend what had just happened. My brother-in-law skillfully negotiated an alternative order and asked that they expedite it.

So now other customers' orders were being delayed while they processed our expedite order.

Serious sushi eaters eat their rolls (*maki*) after the individual pieces. The chefs had chosen to make any rolls we ordered last. As they were self-organizing and burning down our order list, the information that they could not or would not make the California roll had not been forthcoming until the last moment. Hence, the surprise!

So, what is the moral of this story? Iterative, incremental delivery and self-organizing burn-down aren't enough! There has to be some analysis of the requirements and an attempt to understand the customer's true needs—to understand the customer's definition of quality. Based on this, the priorities for the incremental deliveries should be set. There should be a commitment to the customer—a promise made—and it should be honored.

In addition to the quality problem, this restaurant had a broken organizational structure and poor separation of responsibilities. The "Anderson lunch project" should rightly have been the responsibility of the waiter, who should have been playing the project manager role (and maybe the program manager role). The waiter should have analyzed our requirements and understood our priorities. This should have been communicated to the chefs, and the order of our sushi production should have been negotiated against the competing orders at the time. The sushi chefs should have been responsible purely for the production of sushi. They should not have had any project management, program management, or scheduling responsibility. After all, they had no direct contact with the customer, and, as the system's capacity-constrained resource, they should not have been wasting sushi-making capacity trying to do anything else. (Actually, they weren't wasting capacity; they simply weren't doing a job that needed to be done. This was the root cause of the quality problem.)

All of this goes to show that quality is not a given in Japan. They might be great at making cars, but many of those management lessons have not transferred elsewhere. No end of bowing can make amends to a two-year-old whose expectations are disappointed.

On Transition
Initiatives

CHAPTER
16

I LANDED IN THIS AGILE WORLD THROUGH THE GOOD FORTUNE OF BEING INVITED TO LEAD THE USER INTERFACE DESIGN FOR THE LENDING SYSTEM REPLACEMENT PROJECT AT UNITED OVERSEAS BANK (UOB) IN SINGAPORE, IN 1997. THIS WAS A $20 MILLION PROJECT TO BUILD A SINGLE SYSTEM to manage loans for consumer, commercial, and corporate lending that would replace a whole host of disconnected legacy systems. It was widely believed that such a broad scope had never been undertaken by any bank anywhere in the world. UOB was bold and ambitious, and history has shown that boldness, ambition, and managed risk taking have paid off. By one measure or another, UOB is now the first or second largest bank in Southeast Asia.

UOB in 1998 is where the Feature-Driven Development method was born. It grew from the synthesis of Peter Coad's analysis and design techniques with Jeff De Luca's management techniques, within the context of a 50-person, $20 million multi-year project. Jeff De Luca was a systems thinker. He knew his role as project director wasn't to manage the project, but to build a system of people following a process such that they were capable of building the software and delivering it against a schedule and a budget. The solution was created out of circumstances. The method failed to change the wider UOB IT department or to institutionalize it beyond the lifespan of the project leadership—Jeff De Luca and I left in the spring of 2009; Philip Bradley and Stephen Palmer left less than a year later.

My first attempt to use FDD outside of Singapore was in Ireland in 1999. I told that story in chapter 7, in Another Tale of Belonging. Again, it was a one-off solution for a single, larger-scale project. It failed to institutionalize beyond the lifetime of the project and it failed to change how software is developed at Trinity Commerce/Eircomm.

It wasn't until 2000 that I was asked to run a change initiative to influence the behavior of other teams and departments and to change the way that they worked. It was three years before I realized that trying to change to a defined process—a destination that promised a nirvana of productivity and quality—was apt to meet with resistance, and that failure to institutionalize change was the norm rather than the exception. As I finished my first book, *Agile Management for Software Engineering,* I started to turn my attention to what I believed was the real challenge—change management and institutionalization of process improvements.

The Agile movement has a handful of core concepts, two of which are these: We should make progress with imperfect information; and we should build adaptive systems that give feedback and adjust. These concepts are built on the assumption that we cannot know everything we need to know in advance, and that predicting the future is a low-leverage activity. Hence, it was ironic to me that Agile methods were introduced in an arrogant fashion with the assumption that they'd simply be better, without any specific context taken into account. The defined processes outlined in the textbooks should be implemented over a transition period that integrates education, training, practice, and coaching. The method of managing change was traditional. There was no feedback loop or adaptive process for introducing change. Ironically, Agile methods were introduced in an anti-Agile fashion.

This was the core driver for the emergence of the Kanban Method. My work on Kanban can be traced directly back to 2003, as I completed my first treatise on Agile management. So, it is poignant that I conclude this second volume on Agile management with my collected thoughts on transition and the emergence of a truly Agile approach to change management—the Kanban Method.

No More Quality Initiatives

Wednesday, April 27, 2005

• •

The problem with quality initiatives, and their champions, and their change agents, and their sponsors, and their improvement projects, and their quality teams, and process experts, and black belts, and green belts, and funny handshakes, and secret rituals, and coded handbooks, and group hugs, and therapy sessions is, quite simply, that the improvements don't last.

• •

ABOUT SIX MONTHS BACK WE (THE MICROSOFT VISUAL STUDIO TEAM SYSTEM group) hosted a meeting of our customer advisory council. This is a handpicked group of people who help to steer our product. They all know who they are, so I don't need to name names. Many of them read this column regularly. Back then, I was soliciting input for our "formal" methodology. During this session, one of our advisors urged me to avoid giving us "yet another quality initiative. We've tried them all and people are tired!"

And so, I'm sure he'll be delighted to hear me say, "Quality initiatives! Just say no!"

The problem with quality initiatives, and their champions, and their change agents, and their sponsors, and their improvement projects, and their quality teams, and process experts, and black belts, and green belts, and funny hand-shakes, and secret rituals, and coded handbooks, and group hugs, and therapy sessions is, quite simply, that the improvements don't last. When the black belt goes home, and everyone exhales, the regular workforce goes back to their same old behavior. The fix for this is to focus on the culture and make . . .

QUALITY EVERYONE'S BUSINESS!

Simply put:

CONTINUOUS IMPROVEMENT IS EVERYONE'S BUSINESS!

Plain and simple—quality and continuous improvement are everyone's business. It ought to be an everyday thing for every employee. Everyone should understand variation, and specifically, understand how to measure and interpret the variation in their inputs, their rate of input, their working method, their lead time, and their rate of output. Eliminating special-cause variation should be everyone's business, every day. Reducing common-cause variation should be everyone's business, every day. Suggestions for improvement could come from anyone, at any time, and be implemented by a local consensus on the shop floor—no need for a central process improvement group or sanction from an ivory tower full of process priestesses.

That's why in MSF for CMMI* Process Improvement, I've included daily standup meetings to bring up issues and monitor and manage risks and to eliminate special-cause variation and make it everyone's business to do so. That's why we're dropping conformance to plan and conformance to specification in favor of conformance to process and focus on variation reduction. That's why we're encouraging a bottom-up, empowered team, consensus model that allows decentralized decisions to be made quickly. The way to institutionalize continuous improvement across an organization is to make it everyone's business, every day! The way to deliver an Agile process that meets both the original spirit of the software CMM and the letter of the CMMI appraisal model is to distribute the quality responsibility at a grassroots level across the whole organization. Everybody, every day allows quality and agility to walk hand in hand.

● ● ● ● ● ● ● ● ● ● ● ● ● ● ● ● ●

* Looking back, the focus of my work and the slant of this article is perhaps too focused on reduction of variability rather than my contemporary approach to study the demand and classify the risks in the demand—identifying valuable variety while reducing and eliminating undesirable variation in flow.

However, this article does highlight the essence of my approach to change—that it must be cultural and that the focus of any management-led initiative must be to drive cultural change toward a culture of continuous improvement. The workplace and workflow changes will then take care of themselves.

A process-led change initiative, in which the workforce believes that responsibility for change (and improvement) has been delegated to a specific group, such as process engineers or coaches, is likely to lead to an attitude of "improvement isn't my responsibility," and resistance to changes that are seen as being imposed from the outside. It's clear that by the spring of 2005, I'd already realized that the desire to change and make process improvements had to come from within, and that if this was to happen, the focus of change led by management must be cultural.

I wasn't explicitly connecting this with Toyota's *kaizen* cultural approach, but it is clear that my intent was similar. This was, then, a highly differentiated stance against what was common in the Agile community. This is still true in 2012.

Thoughts on Enterprise Agile Transition

Sunday, September 23, 2007

. . . .

One of our execs met some guy sitting in first class,
who persuaded him that if we weren't Agile, we were
being left behind.

. . . .

WHILE AT AGILE 2007, I ATTENDED A DISCOVERY SESSION LED BY PETE BEHRENS. He had assembled three of his clients, who had all taken their teams through enterprise-level transitions to an Agile approach to software development and project management, along with a group of perhaps 60 conference attendees. We sat in small circles of six or so folks per group. The session was divided into talks by each of Pete's clients, some summarization from Pete himself, and group chats, followed by some "show and tell" to the wider room.

My big take-away from the session was that several of the folks in my small group and at least one of the main presenters, gave a summary of their transition to Agile that went something like this:

> One of our execs met some guy sitting in First Class, who persuaded him that if we weren't Agile, we were being left behind. So, when he got back from his trip, he sought me out as a known change agent and asked me to lead a transition to Agile development. I assembled a small team and we learned all there was to know about Agile. (Perhaps we hired a consultant to help us.) Then we made a plan of how to take our teams from the traditional waterfall-style development method we'd been using to an Agile/Scrum approach. Over the next nine months, we executed on that plan, and gradually we saw all the Agile practices adopted across all of our teams. We completed that process three months ago, and now we are Agile!

Does anything about this strike you as ironic?

The way to enterprise agility was to make a big plan up front, based on a top-down, management-led initiative, and to command and control the team to change to an Agile style of working. Then they executed against the plan, and when every task on the plan was completed, they declared that the change was done!

I've been there before, and I struggled to achieve both scale and institutionalization of the changes. My fear with many of the enterprise-scale transitions now taking place is that they will suffer the same fate. When the management that led the change is gone, teams will gradually atrophy back to a traditional way of working. Unless a fundamental change in the organizational culture is achieved, and the new culture is institutionalized from top to bottom in the organization, I fear that the benefits of agility will be short-lived. Agile may be labeled as just another management fad.

Institutionalization of Culture versus Prescribed Process Change

Sunday, September 23, 2007

• •

In the past, I've been an aggressive change agent. The CMO at Sprint PCS described me as "hard driving." That approach achieved localized results, but equally, it made me unpopular with some peers.

• •

WHILE SUCCESS STORIES ARE ALL VERY WELL, WE IN THE AGILE COMMUNITY know that we learn more from our mistakes and challenges than from our successes. Hence, I thought it might be more useful for my readers to hear some of my frustrations from my first year leading the software engineering team at Corbis than simply to read my self-congratulations from last week.

I blogged before that I took a different approach on joining Corbis. I went after the culture rather than leading change to a specific prescriptive method, such as Feature-Driven Development. I did this because I wanted to achieve organization-level change and institutionalization of the results. Previously, I had enjoyed localized success with my immediate development team or project but I had not managed an enterprise-level adoption, nor had any changes survived my departure by more than a year—at both Sprint PCS and Motorola, earlier in the decade. I was afraid of the J-curve effect that driving home a specific prescribed process change might have at Corbis. I didn't want the downside of change to cause panic and a knee-jerk reaction back to the comfort zone of the status quo. So I opted for the long game of cultural change.

I'm delighted with the results, and the cultural change, while hard to measure objectively, there is wide recognition of the improvements from anecdotal evidence. However, I remain frustrated by some of the results. I've seen a lot of code released and a lot of successful software releases. I've seen very high quality with very few defects escaping into our production environments, but I haven't yet seen hyper-productivity.

Jeff Sutherland talks a lot about how difficult it is to achieve hyper-productivity. He talks about how few Scrum teams have ever achieved it. (He defines this as performance that is at least four times higher than might be considered normal.) He holds up the team that created Borland Quattro Pro as one of the highest performing teams of all time—a team that realized at least a ten-fold productivity gain. At Agile 2007, I spent some time with Jeff and shared with him the data from the Device Management project that Daniel Vacanti and I led at Motorola in 2004. The data compares favorably with the Borland Quattro Pro project. Even data from lesser-performing FDD projects (such as the Singapore project—though it was a much larger project), and others from my team at Sprint, and some others from FDD teams I've met along the way, tends to indicate hyper-productive results, and often with incredibly high-quality, trustworthy data. Hence, over the years I've been used to achieving hyper-productivity with my teams.

So what's up at 710 Second Avenue in Seattle? Why no hyper-productivity?

My guess on this is that with cultural change, and institutionalization and what goes with it—a high degree of autonomy, delegation, empowerment, and self-organization—it takes a lot longer to achieve hyper-productive results. The advantage is that change is achieved with a much lower degree of resistance, with very low attrition and change-driven churn in personnel. The changes, we hope, will stick better and survive future management changes and the general churn in projects and personnel over the coming years.*

In the past, I've been an aggressive change agent. The CMO at Sprint PCS described me as "hard driving." That approach achieved localized results, but equally, it made me unpopular with some peers. It may also have left a residue of resentment toward management-led change among the (admittedly few) skeptical individual contributors. Folks acquiesced and played along with the rules of FDD while I was there, and set the appropriate expectations. They enjoyed the fruits of success from projects delivered on time and with high quality. However, when management wasn't there to enforce the rules anymore, they fell back into the same old ways of working. The right way is to let the team make its own changes. Through ownership of change, the seeds of long-term support and advocacy are born.

So while I wait for hyper-productivity to emerge or evolve through a series of *kaizen* events, I'm not unhappy. I set out to change a culture and to achieve an institutionalization of those changes, which will eventually survive me and the rest of the management team. I've delivered on that goal, but darn it, I miss the kick out of seeing some hyper-productivity.

*Ironically, the improvements at Corbis survived only about nine months after I left in 2008. I attribute it mainly to the CIO's decision to reduce operations reviews to quarterly, and then to drop them altogether. These actions signaled a change in culture. They signaled that objective, data-driven operations were no longer valued, and that senior management didn't much care about capability or process improvement. There was no longer any respect for the people doing the work or the methods they were employing to do the work. It serves to highlight that corporate culture emerges from the sum of management actions and decisions. It takes only a few decisions and actions to significantly change a culture (for better or worse.)

Focus on Organizational Maturity (not Appraisals)

Friday, October 17, 2008

• •

It is being said that I suggested organizations must seek a CMMI appraisal in order to deliver on their goal of enterprise agility.

• •

I'VE BEEN GREATLY ENCOURAGED BY THE FEEDBACK FROM MY AGILE 2008 MAIN Stage presentation. However, something I've heard elsewhere has resurfaced. My assertion that a high level of organizational maturity is required to achieve institutionalized enterprise-scale Agile adoption is being misinterpreted. It is being said that I suggested organizations must seek a CMMI appraisal in order to deliver on their goal of enterprise agility.

Reinout van Rees[1] wrote:

> In David's opinion, kanban is the method that can help us achieve both agility and high maturity. It will push us forward. It creates a cultural shift. A shift that allows some teams to reach CMMI [maturity] Level 4

I want to be categorical that I am not saying that appraisal is necessary. What I am saying is that the CMMI is an existing model for organizational maturity. It is a model with 20 years of learning and iteration built in to it. The people in charge of it are still learning and still iterating. My observation of real teams adopting Agile is that they appear to more or less follow the CMMI's model of organizational maturity. In other words, Level 2 practices appear first, then Level 3, and so on. My conclusion from this observation is that the CMMI model for organizational maturity is more or less correct—close enough to be good enough.

What I've observed with teams pursuing a Kanban approach is that it creates a culture suitable for the generic practices in the CMMI model that lead to high maturity. In addition, Kanban appears to create an appropriate framework for the practices in Levels 2 and 3 of the CMMI model to emerge naturally/

---

1. Great thinkwork about "agile": Presentation by David Anderson http://reinout.vanrees.org/weblog/2008/09/01/agile-david-anderson.html

organically/spontaneously without the need for a CMMI process initiative. This is a significant win because a common anti-pattern with CMMI is "management by objectives" where the objective is to get an appraisal at a specific level.

Hence, what I am saying is that if we know that a high level of organizational maturity is a key indicator of success with enterprise-scale adoption, it makes sense to pursue organizational maturity as a driver to success. If pursuing organizational maturity is a goal, we need not re-invent the wheel. We can follow the model that the Software Engineering Institute has provided us. Getting a CMMI appraisal is not part of the message at all. An appraisal might make sense if you are in a regulated industry and have a business driver for an appraisal, such as competing for government contracts. However, appraisal does not enter my message concerning success with enterprise-scale adoption. The message is:

A high level of organizational maturity appears to be essential to successful enterprise-scale adoption. There is an existing model for organizational maturity that appears to be broadly correct. That model is CMMI.

The Relevance of Level 4

Sunday, November 2, 2008

••••••••••••••••••••••••••••••••••••

CMMI Maturity Level 4 is often thought of like Nebraska or Kansas—it's the flyover territory of CMMI.

••••••••••••••••••••••••••••••••••••

CMMI MATURITY LEVEL 4 IS OFTEN THOUGHT OF LIKE NEBRASKA OR KANSAS—it's the flyover territory of CMMI. The big offshore outsource companies often think of Level 4 as something that they can skip—jumping from Level 3 to Level 5. After all, there are only four process areas, two in Level 4 and two in Level 5.

When I was at Microsoft, working on MSF for CMMI® Process Improvement, we talked about the future prospect of an enhanced edition that would provide full coverage of Levels 4 and 5. There was no market demand for a Level 4 solution. Our market research was telling us that there was a market for a Level 3 solution—the one we produced—aimed at the government contracting market in North America and the ISO 9000 compliance market in South America. We also knew that there was a market for a Level 5 template for Team Foundation Server (TFS)—mostly aimed at the offshore outsourcing companies. Level 4 just didn't come into our plans. It was flyover territory. It seemed that no one does Level 4. If you look at the list of CMMI-appraised firms, there are very few at Level 4. So, why am I suddenly a big advocate of Level 4?

Well, from discussions with clients and potential clients in America and Europe, it seems our clients need to have the equivalent of Level 4 organizational maturity in order to meet their business goals and strategic objectives. They don't need to be an optimizing organization at Level 5—that would be icing on the cake. But they do need to be predictable. They want to have strong delivery with low variability. They want to be proactive and drive down cycle times using objective quantitative management. They need all of this to deliver on business goals within the tight financial controls and corporate governance that they now find themselves under. They need to be the equivalent of Level 4.

The real problem is that typical Agile methods can take them only to Level 3. So Agile isn't enough. That's an opportunity for my consulting firm. Our experience in creating cultures that drive toward high maturity (Levels 4 and 5) while implementing Lean and Agile techniques is still fairly unique. We help clients reconfigure their organizational culture to enable a high-maturity organization to emerge while still gaining all the benefits of Agile and Lean methods.

CMMI Maturity Level 4 has real business relevance. Businesses that achieve it will achieve their goals, hit their numbers, and delight customers, shareholders, and employees. Getting to Level 5 will allow a firm to become ever more competitive and to dominate their market. But for many firms, the need to achieve the equivalent of Level 4 maturity is a business imperative—now! Anything less will leave all stakeholders dissatisfied.

My Approach at Corbis

Monday, November 13, 2006

• •

What am I hoping to achieve with this new *kaizen* culture? Gradual sustained improvement! Rather than a big bang to a defined and prescribed methodology, I'd like to morph the system gradually. . . .

• •

I'VE BEEN AT CORBIS ALMOST TWO MONTHS NOW. ALMOST 60 OF MY FIRST 90 days are gone. So, what am I doing? People have been asking me what my approach is. Am I implementing Feature-Driven Development across the organization? Or perhaps an enterprise-wide Scrum initiative? Or even MSF with Visual Studio Team System? And if so, which flavor, Agile or CMMI?

These are all good questions. And the answer is none of the above. Why not?

I personally feel that if I try to prescribe a method and then drive a big change initiative to train everyone on using it, and I use managers and the small process engineering group as enforcers, I will fail miserably. True, a younger version of me, six years ago, for example, would have chosen to drive an FDD prescription as a defined process to be followed rigorously. But not anymore. I will fail if I try this again. Why?

There are two reasons. Those who watched the video of my exit interview with Microsoft's Channel 9 know that I am committed to changing Corbis engineering. My goal is to increase productivity and quality—with reduced variation in the process and output—and without losing any of the loyal staff along the way. This is a real challenge. In a recent exchange with Ken Schwaber in the Scrum Development Yahoo! group, Ken advised that some of the enterprise Scrum implementations have resulted in up to a 30 percent staff turnover. Ouch! I can't afford for that to happen.

I also have commitments to deliver some large projects along the way. I cannot *not* deliver them. I can't allow them to be late or of poor quality because we happen to be retraining the workforce in a new method. In other words, the J-curve effect would be too deep and too long. Long, deep J-curves are a recipe for getting junior executives fired. And I don't want that to happen.

So, what am I doing?

I'm creating a *kaizen* culture, encouraging grassroots, shop-floor improvement suggestions within a structured objective framework of management metrics. We're implementing work-item tracking on Team Foundation Server, reporting with SQL Reporting Services, and transparency through the use of the Team Portal and by deploying TeamLook (the MS Outlook client for TFS).

To encourage the growth of the right culture, we're starting monthly all-hands operations reviews, similar to those I describe in Chapter 14 of my book, *Agile Management for Software Engineering—Applying the Theory of Constraints for Business Results*.* My direct reports have been working to devise the metrics they'll be reporting at the ops review. The first one is on December 7, for the month of November. Improvement suggestions resulting from the operations review will be assigned to managers as *kaizen* action-item events and followed up in future ops reviews. Everyone on the staff will learn that management is committed to making them more successful and getting them what they need to improve productivity and quality with reduced variation.

We've also been retooling our sustaining process for minor enhancements and bug fixes. We've been doing that by implementing a kanban system for processing change requests. The implementation is a little different from the one implemented at Microsoft's XIT department in 2004. The main difference is that there aren't dedicated team members for sustaining engineering, but our commitment is to provide a sustained level of service and a guarantee of capacity. This was tricky to achieve, but the kanban system really helped. I hope that next year I'll be able to report favorable results.

What am I hoping to achieve with this new *kaizen* culture? Gradual, sustained improvement! Rather than a big bang to a defined and prescribed methodology, I'd like to morph the system gradually, identifying bottlenecks and areas of extreme variation and unreliability, and designing improvement programs for them, one by one.

This means that I have no idea how it is going to turn out, but I'm sure that we'll all have fun getting there. Corbis was already a great place to work, underscored by the large number of loyal, long-serving staff. By introducing *kaizen* to the culture, I hope to enhance that even more.

● ● ● ● ● ● ● ● ● ● ● ● ● ● ● ● ●

*Coincidentally, chapter 14 of the first edition of *Kanban: Successful Evolutionary Change for Your Technology Business* was also about operations reviews.

• • • • • • • • • • • • • • • • • •

**Revisiting this blog post more than five years later, it is interesting to note a couple of strong themes—the emphasis on cultural change led by setting an expectation of objectivity and openness, and the recognition that a kaizen culture would produce emergent results. There would be no defined destination process, and as a result, no testable definition of done. I was building a truly evolutionary organization that would be robust and resilient to changes in the business it was supporting—and in its own organizational structure and operations. It was this organization that created the Kanban Method that we know today. Kanban was a gift that emerged from a focus on culture rather than a managed transition to a defined process definition. All that was left for me to do was to write it down and share it with the world.

Six Month Review

Tuesday, March 20, 2007

• •

So what is the lesson to be learned here? I think it is very simple—there is more than one way to achieve agility. Practices from eXtreme Programming and Scrum are just possible approaches.

• •

YESTERDAY, MARCH 19, WAS MY SIX-MONTH ANNIVERSARY AS SENIOR DIRECTOR of Software Engineering at Corbis. This gave me occasion to pause and take stock of how I'm doing. Tomorrow, Wednesday, March 21, we will make the thirteenth software release to production since I joined the team. That's an average of a release to production every 9.5 business days since mid-September, 2006. In addition, we haven't had an escaped defect to production since the beginning of the year. I'll repeat that for emphasis—zero escaped defects since January 1, 2007. I'm very impressed with the team I inherited, and this track record is something they can all be proud of.

If we were writing a set of acceptance tests for an Agile process, those tests might ask:

❐ Is software being released to production every two weeks or less?

❐ Is the quality of that software high?

❐ Is the technology organization able to respond to changes in the business rapidly?

On all three of these questions, my team would pass. What is really interesting is that we aren't doing many things that would be recognized as "Agile." If there were a set of white-box tests to test for Agile practices, such as:

❐ Are they pair programming?

❐ Are they using test-driven development?

❐ Are releases planned in two-week iterations?

❐ Is there continuous integration running?

❐ Are they reporting burn-down?

❐ And so on.

On these tests, we would fail. We are using a daily standup and we do track progress on a white board. But apart from that, our process would not be recognizable as Agile.

So, what is the lesson to be learned here? I think it is very simple—there is more than one way to achieve agility. Practices from eXtreme Programming and Scrum are just possible approaches. What many in the Agile community see as "Agile" is just a recipe that works in some situations. There are many other approaches that can be just as agile (deliberate use of lower case) and that might be more appropriate to your situation. By sticking to the principles in my Recipe For Success[1]—focus on quality, reduce work-in-progress, balance demand against capacity, and prioritize—my team has been able to deliver a continuous flow of business value that enables Corbis to be more agile in the marketplace.

Although there is a lot more to be done, my career at Corbis is off to a good start. The challenge now is to take our success with software maintenance and apply it to major projects in our portfolio. As a result, I'm even more excited about the next six months than I was about the first.

1. Anderson, David J., *Kanban: Successful Evolutionary Change for Your Technology Business*, Blue Hole Press, 2010, chapter 3.

One Year Anniversary

Monday, September 17, 2007

. . . .

It's amazing how a values-driven organization can mature toward a self-governing culture enabled by transparency and high levels of trust.

. . . .

TODAY WAS MY ONE-YEAR ANNIVERSARY IN MY JOB AS SENIOR DIRECTOR OF Software Engineering at Corbis. It's been an interesting year. Although we put out 34 releases of new software across a range of IT systems and external facing systems, such as corbis.com, with barely a handful of defects, the really interesting part of the job has been watching the cultural change and the growing maturity of the whole IT organization. It's amazing how a values-driven organization can mature toward a self-governing culture enabled by transparency and high levels of trust. The results are greater productivity and exceptional quality.

After a year, I'm having more fun than ever. It's fascinating working with a team as we gradually change from a traditional waterfall approach to our IT projects, and introduce a more collaborative form of working. There is a lot more work to be done, and I'm looking forward to pushing the boundaries with enterprise Agile and Lean methods. Meanwhile, our Kanban ideas are really catching on, with over 130 members in the Yahoo! group and close to 300 messages already this month. We are seeing some bigger companies like Yahoo! looking into it and sharing their progress with other group members. It's so energizing to see our ideas catching on, both within our own company, and across the wider software-development community around the world.

. . . .

As this book goes to press, the Yahoo! group has over 2,250 members, a newer LinkedIn group has over 1,100, more than 900 people attended Kanban-specific conferences in 2011, and more than 10,000 copies of my book, *Kanban: Successful Evolutionary Change for Your Technology Business*, have been sold. Kanban training is offered in 26 countries. There is Lean Kanban University, offering standards in the definition of Kanban, and both a training curriculum and quality trainers via 20 member firms in 11 countries. I am aware of large-scale implementations at brand-name businesses on five continents. It's been wonderful to nurture and guide the adoption and acceptance of Kanban and to watch in wonder as it emerges as a major influence on how knowledge work will be conducted this century.

References

TBD

Index

domain modeling, 260, 329, 349, 357

Drucker, Peter, 8, 18, 32, 65-66, 75-84, 86-88, 290, 293,361

Drum-Buffer-Rope system, 142-3

E

Edwards, Martin, 48

enabler, 218, 237, 259-60

enemy, common, 185, 213-14, 222-4

enterprise-scale adoption, 413

estimation, 10, 116, 252, 278, 280-2

evolutionary approach, 25, 126, 139, 153, 259, 261, 271

expedite request, 320

exploit, 147, 150, 177-8, 380

eXtreme Programming, 14, 107, 120, 185, 194-6, 198, 200, 258, 261, 340, 370, 419-20

F

failure, 7, 17, 28-31, 45-6, 57-60, 104, 112, 115, 156, 170, 180, 285, 288, 293, 310

FDD (Feature-Driven Development), 19, 35-7, 65, 88, 121-2, 125, 156, 185, 190, 256, 294, 322-3, 396-7, 403, 409-10

FDD projects, typical, 260

feature classification, 377-8, 380

Feature-Driven Development. *see* FDD

features, commodity, 347-8, 377-8

Ferguson, Alex, 7, 47-9, 242

Five Focusing Steps, 126, 135, 147-8, 150, 152-3, 157, 291

flow, 1, 14, 18, 21, 32, 75, 126, 136, 151, 161-2, 176, 290-1, 298, 319-20, 387

   cumulative, 114-16, 119

   smooth, 319-20

flow model, 291

flow systems, 333-5

forecasts, 318, 334

Fowler, Martin, 11, 195, 272, 315-18

G

genchi gembutsu, 11, 341-4, 346

Goldratt, Eli, 1, 8-9, 125-7, 128-29, 130-2, 134-8, 140, 142, 144, 146, 148, 150, 152-4, 156-7, 189, 239

Greenland Norse, 262-4

group accountability, 288, 293, 296, 300, 308

group project manager, 130

H

Hedgehog Concept, 89-90, 121-2

high-trust culture, 46, 258, 262, 264

Human Resources (HR), 231, 236-41

HR Myths, 10, 228-9, 231, 233, 235-9, 241, 243

HR policies, 234, 238-9

hyper-productivity, 409-10

I

IBM, 34, 67, 236-7

Ideal State, 102-5, 134

Immelman, Ray, 185, 197-8, 208, 210, 213-14, 218, 239

Immelman's model, 198-9, 223, 239

improvement, continuous, 103, 109-10, 112, 296, 405-6

improvement projects, 405

information, imperfect, 10, 256-8, 329-30, 404

initiative, management-led, 406-7

input queue, 382-3, 387, 392-3

inventory, 156-7, 226, 278, 315-18, 321, 324, 326, 345

issue log, 103, 115-17, 291

iteration length, 117, 154-5

iteration plan, 117

iterations, 58, 82, 88, 95, 114-17, 128, 154-5, 244, 259-60, 269, 283-4, 296, 299, 306-7, 331-2

K

kanban, 136, 185, 209, 313, 412

Kanban community, 66, 91-2, 102, 136, 185, 205, 221, 247, 310

Kanban Method, 20, 118, 123, 126, 132, 149, 152-3, 205, 261, 264, 298-9, 301, 330, 392, 404

kanban system design, 139, 330, 383

Kerievsky, Joshua, 82, 200, 271, 313

L

LAAAM (Lightweight Alternative Architecture Assessment Method), 345

leadership, 1, 7, 14-15, 17, 25-6, 28, 30, 32, 34, 36-8, 40, 42-4, 46, 48, 123

lean, 45-6, 185, 215, 317-18, 387

Lean, 1, 8, 11, 14, 17-18, 25, 122, 135-6, 177-8, 185, 247, 313-16, 318-20, 324-6, 362

Lean concept, 313, 344, 346

Lean Kanban University, 274, 421, 437

lean manufacturing, 18, 136, 315-17, 339, 350-1

Lean Product Development, 314, 330

lean project management, 244

Lean Software Development, 20, 313-14, 328, 342, 346, 348, 350, 357-8, 362, 399

Lightweight Alternative Architecture Assessment Method (LAAAM), 345

line managers, 34, 60, 99, 141, 186, 237, 390

M

managed change initiatives, 110

Management Accounting, 239

management misdirection, 7, 57, 59-60, 84

  classic, 58-9

management of process variation, 65-6

Managing, 61, 68, 77, 120, 324, 328

market risk, 11, 369, 375-8, 380-1, 383

market segmentation, 375, 378, 380

maturity, high, 412, 414

Maturity Levels, 106-7, 109-10, 112-13

measure capability, 147-8

measure developers, 85

measure individuals, 65, 68, 85-6

measurements, 53, 57, 63, 100, 267

  individual, 8, 57, 84-6

meritocracies, 230, 232

Microsoft, 19, 92, 106, 113, 149, 175, 188, 213-15, 220, 227-8, 236-7, 239-40, 250-2, 312, 397-8

Moment-Interval, 352-4, 373

morale problem, 68

motivation, 30-1, 86, 159-60, 185, 197, 232, 251, 256, 271

MSF for Agile Software Development, 107, 110, 118

Mulhern, Dan, 217-19

N

Nextel, 224, 367-8

O

objectives, management by, 77, 413

option-based decision, real, 371

option theory, real, 371, 381

options, real, 11, 370-2, 375

organizational structure, 12, 45, 84, 197, 387-8, 390, 392, 394, 396-9, 418

variation, 66, 93-5, 100-3, 109, 112, 117, 135-6, 148, 151-2, 278-9, 282, 291-2, 347, 363-5, 405-6
  bad, 347
  cause, 100
  measures, 113
  natural, 107, 109, 117, 147, 150-1
  observed, 94
  reduced, 65-6, 92, 94, 416-17
velocity, 113, 117, 119, 192, 278-9, 351
Visual Studio Team System, 416

W

waste, 45-6, 133, 135, 252, 256-8, 271, 277-8, 280, 317, 319-20, 396-8
WIP, 114-15, 161, 316-20, 323-4, 326-7, 420
WIP levels, 114
work-in-progress, 114-15, 161, 316-20, 323-4, 326-7, 420
workflow, 8, 133-5, 139, 141, 148, 152, 161, 298, 340, 353, 382-3, 394

About the Author

David J. Anderson is a thought leader in managing effective technology development. He leads a consulting, training, and publishing business dedicated to developing, promoting, and implementing sustainable evolutionary approaches for management of knowledge workers.

He has 30 years' experience in the high technology industry, starting with computer games in the early 1980s. He has led software teams delivering superior productivity and quality using innovative agile methods at large companies such as Sprint and Motorola. David is credited with the first implementation of a kanban system for software development, in 2005. Later he created the Kanban Method, a set of core principles and practices that is the basis for accredited Kanban training curricula worldwide.

David is the author of two earlier books, *Agile Management for Software Engineering: Applying the Theory of Constraints for Business Results,* and *Kanban: Successful Evolutionary Change for Your Technology Business.*

David J. Anderson was ranked #5 in the Top 20 Most Influential Agile People (2012) and his book, *Kanban: Successful Evolutionary Change for Your Technology Business,* was ranked #4 in the Top 100 Agile Books (2011).

David is a founder of the Lean Kanban University, a business dedicated to assuring quality of training in Lean and Kanban throughout the world.

He is based in Sequim, Washington, USA.

Get in Touch

Sign up for David J. Anderson's monthly newsletter or the Kanban Weekly Roundup at http://djaa.com.

To request training or consulting in your area, please contact sales@djaa.com. Upcoming classes are also listed at http://www.djaa.com/events.

Join discussions about Agile and Lean development at the Yahoo Group for Agile Management or the Yahoo Group for Kanban development.

- ❏ http://finance.groups.yahoo.com/group/kanbandev
- ❏ http://finance.groups.yahoo.com/group/agilemanagement

Connect with the Lean and Kanban communities at the Limited WIP Society's website, http://limitedwipsociety.org.